Lecture Notes in Mathematics   1587

Editors:
A. Dold, Heidelberg
B. Eckmann, Zürich
F. Takens, Groningen

Subseries: Mathematisches Institut der Universität
und Max-Planck-Institut für Mathematik, Bonn - vol. 20

Adviser: F. Hirzebruch

Nanhua Xi

Representations
of Affine Hecke Algebras

Springer-Verlag

Berlin Heidelberg New York
London Paris Tokyo
Hong Kong Barcelona
Budapest

Author

Nanhua Xi
Institute of Mathematics
Academia Sinica
Beijing 100080, China

Mathematics Subject Classification (1991): 22E50, 20G05, 19L47, 19K33, 18F25, 20F55, 16S80

ISBN 3-540-58389-0 Springer-Verlag Berlin Heidelberg New York

CIP-Data applied for

© Springer-Verlag Berlin Heidelberg 1994
Printed in Germany

Typesetting: Camera ready by author
SPIN: 10130124   46/3140-543210 - Printed on acid-free paper

Introduction

Let K be a p-adic field with finite residue class field of q elements. Let $\mathcal{G}$ be a connected split reductive group over K with connected center. Let $\mathcal{I}$ be an Iwahori subgroup of $\mathcal{G}$ and $\mathcal{T}$ the 'diagonal' subgroup of $\mathcal{I}$ (in a suitable sense). The group $N_{\mathcal{G}}(\mathcal{T})/\mathcal{T}$ (here $N_{\mathcal{G}}(\mathcal{T})$ is the normalizer of $\mathcal{T}$ in $\mathcal{G}$) is an extended affine Weyl group W (i.e. $W = \Omega \ltimes W'$ for certain abelian group Ω and for certain affine Weyl group W'). It is known that $\mathcal{G} = \bigcup_{w \in W} \mathcal{I} w \mathcal{I}$ and one can define an interesting associative ring structure on the free abelian group H_q with basis $\mathcal{I} w \mathcal{I}$, $w \in W$ (see [IM]). The ring H_q is an affine Hecke ring and we call $\mathbf{H}_q = H_q \otimes \mathbb{C}$ an affine Hecke algebra. According to Borel [Bo1] and Matsumoto [M], the category of admissible complex representations of G which have nonzero vectors fixed by $\mathcal{I}$ is equivalent to the category of finite dimensional representations (over $\mathbb{C}$) of $\mathbf{H}_q$. Thus an interesting part of the study of representations of p-adic groups can be reduced to that of affine Hecke algebras.

According to a conjecture of Langlands (see [La]) the irreducible complex representations of $\mathcal{G}$ should be essentially parametrized by the representations of the Galois group $\mathrm{Gal}(\bar{K}/K)$ into the complex dual group $\mathcal{G}^*(\mathbb{C})$ of $\mathcal{G}$ (in the sense of [La]): $\mathrm{Gal}(\bar{K}/K) \to \mathcal{G}^*(\mathbb{C})$.

Let Γ be the quotient group of $\mathrm{Gal}(\bar{K}/K)$ corresponding to the maximal tamely ramified extension of K. The group Γ has the generators F (Frobenius) and M (Monodromy), subject to the relation $FMF^{-1} = M^q$. According to the conjecture, the irreducible complex representations of $\mathcal{G}$ which have nonzero vectors fixed by the Iwahori group $\mathcal{I}$ should be essentially parametrized by the homomorphisms $\Gamma \to \mathcal{G}^*(\mathbb{C})$. More exactly, Langlands' original conjecture says that the representations should roughly be parametrized by the conjugacy classes of semisimple elements in $\mathcal{G}^*(\mathbb{C})$. A later refinement of the conjecture, due independently to Deligne and Langlands, adds nilpotent elements in the picture. Thus the representations considered should be essentially parametrized by the conjugacy classes of the pair (s, N) such that $\mathrm{Ad}(s)N = qN$, where s is a semisimple element of $\mathcal{G}^*(\mathbb{C})$, N is a nilpotent element in the Lie algebra $\mathbf{g}$ of $\mathcal{G}^*(\mathbb{C})$, and we say two pairs (s, N), (s', N') are conjugate if $s' = gsg^{-1}$, $N' = \mathrm{Ad}(g)N$ for some $g \in G$. For group $GL_n(K)$ this was proved by Berstein and Zelevinsky [BZ], [Z]. For general case, Lusztig (see [L4]) added a third ingredient to (s, N), namely an irreducible representation ρ of the group $A(s, N) = C_G(s) \cap C_G(N)/(C_G(s) \cap C_G(N))^0$ (here $G = \mathcal{G}^*(\mathbb{C})$ and $C_G(\cdot)$ denotes the centralizer in G) appearing in the representation of the group $A(s, N)$ on the total complex coefficient homology group of $\mathcal{B}_N^s$, here $\mathcal{B}_N^s$ is the variety of Borel subalgebras of $\mathbf{g}$ containing N and fixed by $\mathrm{Ad}(s)$.

Now the category of admissible complex representations of $\mathcal{G}$ which have nonzero vectors fixed by $\mathcal{I}$ is equivalent to the category of finite dimensional representations (over $\mathbb{C}$) of the Hecke algebra $\mathbf{H}_q$ with respect to the Iwahori group $\mathcal{I}$ (see [Bo1, M]). Therefore the conjecture can be stated as

(∗) The irreducible representations of $\mathbf{H}_q$ are naturally 1-1 correspondence with the conjugacy classes of triples (s, N, ρ) as above.

The conjecture (∗) was proved by Kazhdan and Lusztig in [KL4]. Actually they proved that (∗) is true when q is not a root of 1 (one can define $\mathbf{H}_q$ for arbitrary $q \in \mathbb{C}^*$). In [G1] Ginsburg also announced a proof, but the proof contains some errors since the main result is not correct as stated, see [KL4, p.155]. However the work [G1] contains some very interesting ideas. Combining [KL4] and [G1] we can prove that (∗) is true if the order of q is not too small (see chapter 6, Theorem 6.6, actually we get more). In chapter 7 we shall show that (∗) is not true if q is a root of 1 of certain orders. It is expected that (∗) is not true only when q is one of those roots of 1 (see [L17]).

In this book we also show that cells in affine Weyl groups are interesting to understand representations of affine Hecke algebras.

Now we explain some details of the book.

In chapter 1 we give the definitions of Coxeter groups and of Hecke algebras. We also recollect some definitions and results in [KL1, L6], which will be needed later. In chapter 2 we give the definitions of extended affine Weyl groups and of affine Hecke algebras, and recall some results on cells in affine Weyl groups. Following Berstein, the center of an affine Hecke algebra is explicitly described. In chapter 3 we describe the lowest generalized two-sided cell of an affine Weyl group (Theorem 3.22). Naturally in chapter 4 we generalize Kato's result on q-analogue of weight multiplicity (see [Ka2]).

In chapter 5 we recall some work on Deligne-Langlands conjecture for Hecke algebras by Ginsburg [G1-G2], Kazhdan and Lusztig [KL4]. We give some discussions to the standard modules (in the sense of [KL4]). For type A_n it is not difficult to determine the dimensions of standard modules. We also state two conjectures, one is concerned with the based rings of cells in affine Weyl groups, and another is for simple modules of affine Hecke algebras with two parameters, which is an analogue of the conjecture (∗). In chapter 6 we introduce an equivalence relations in $T \times \mathbb{C}^*$, where T is a maximal torus of a connected reductive group over $\mathbb{C}$. Combining some properties of the equivalence relation, results of Ginzburg and of Kazhdan & Lusztig in chapter 5, we prove that (∗) is true when the order of q is not too small (Theorem 6.6). In chapter 7 we show that if q is a root of 1 of certain orders the conjecture (∗) is not true by using some results in [Ka2] and in chapter 6.

In chapter 8 we unify the definitions of principal series representations in [M, 4.1.5; L2, 8.11] by means of two-sided cells of an affine Weyl group and also give some discussions to the representations. In chapter 9 we are interested in relations among affine Hecke algebras of the same root system. In chapter 10 we give some discussions to certain remarkable quotient algebras of $\mathbf{H}_q$.

The chapters 11 and 12 are based on preprints "The based rings of cells in affine Weyl groups of type $\tilde{G}_2$, $\tilde{B}_2$" and "Some simple modules of affine Hecke algebras" respectively. In chapter 11 we verify the conjecture in [L14] for cells in affine Weyl groups of type $\tilde{G}_2$, $\tilde{B}_2$. In chapetr 12 we show that the conjecture in [L14] is true for the second highest two-sided cell in an affine Weyl group. Once we know the structures of the based rings we can know the structures of the corresponding standard $\mathbf{H}_q$-modules. The explicit knowledge of based rings provides a way to compute the dimensions of simple $\mathbf{H}_q$-modules and their multiplicities in standard modules, also can be used to classify the simple $\mathbf{H}_q$-modules even though q is a root of 1. In chapter 11 we work out the dimensions of simple $\mathbf{H}_q$-modules for type $\tilde{A}_2$. An immediate consequence is that for type A_2 we see $\mathbf{H}_q \not\cong \mathbf{H}_1 = \mathbb{C}[W]$ whenever q is not equal to 1. This leads to several questions.

I would like to thank Professor T.A. Springer for some helpful conversations. I am grateful to the referees for helpful comments. I wish to thank Professor J. Shi for several helpful comments. I am indebted to Ms. D. Baeumer for the helps in preparing the $\mathcal{A}_{\mathcal{M}}S$-T$_{\mathrm{E}}$X file of the book. Particular thanks are due to the series editors of the Lecture Notes in Mathematics for kind and helpful correspondences. To my pleasure, after the acceptance Professor F. Hirzebruch kindly permits to publish the book as one volume of the Bonn subseries of Lecture Notes in Mathematics.

This work was done during my visit at the Institute for Advanced Study, Princeton, 1991-92, and during my visit at Max-Planck-Institut für Mathematik, Bonn, 1992-93 and December 1993 - September 1994. I acknowledge with thanks the NSF support (Grant DMS-9100383) at the IAS, I am grateful to MPIM for financial support.

Contents

1. Hecke Algebras

In this chapter we give the definitions of extended Coxeter groups and of their Hecke algebras and provide a few examples. Some definitions (such as these of Kazhdan-Lusztig polynomials and cells) and results in [KL1, L6] are recalled. We also show how to apply the definitions in [L6], which are generalizations of those in [KL1]. Several questions are proposed. We refer to [B, Hu] for more details about Coxeter groups and their Hecke algebras.

1.1. Basic definitions. A Coxeter group is a group W' which possesses a set $S = \{s_i\}_{i \in I}$ of generators subject to the relations

$$s_i^2 = 1, \qquad (s_i s_j)^{m_{ij}} = 1 \ (i \neq j),$$

where $m_{ij} \in \{2, 3, 4, ..., \infty\}$. We also write m_{st} for m_{ij} when $s = s_i$ and $t = s_j$.

We call (W', S) a Coxeter system and S the set of distinguished generators or the set of simple reflections. Let l be the length function of W' and $\leq$ denote the usual partial order in W'.

In Lie theory we often need to consider extended Coxeter groups. If a group Ω acts on a Coxeter system (W', S), we then define a group structure on $W = \Omega \ltimes W'$ by $(\omega_1, w_1)(\omega_2, w_2) = (\omega_1 \omega_2, \omega_2^{-1}(w_1) w_2)$. The group W is called an extended Coxeter group. For convenience we also call (W, S) an extended Coxeter group and (W', S) a Coxeter group. The length function l can be extended to W by defining $l(\omega w) = l(w)$, and the partial order $\leq$ can be extended to W by defining $\omega w \leq \omega' u$ if and only if $\omega = \omega'$, $w \leq u$, where $\omega', \omega \in \Omega$ and $w, u \in W'$. We denote the extensions again by l and $\leq$ respectively.

Let $\mathbf{q}_s^{\frac{1}{2}}$, $s \in S$ be indeterminates. We assume that $\mathbf{q}_s^{\frac{1}{2}} = \mathbf{q}_t^{\frac{1}{2}}$ if and only if s, t are conjugate in W. Let $\mathcal{A} = \mathbb{Z}[\mathbf{q}_s^{\frac{1}{2}}, \mathbf{q}_s^{-\frac{1}{2}}]_{s \in S}$ be the ring of all Laurant polynomials in $\mathbf{q}_s^{\frac{1}{2}}$, $s \in S$ with integer coefficients. The (generic) Hecke algebra $\mathcal{H}$ (over $\mathcal{A}$) of W is an associative $\mathcal{A}$-algebra. As an $\mathcal{A}$-module, $\mathcal{H}$ is free with a basis T_w, $w \in W$, and multiplication laws are

(1.1.1) $(T_s - \mathbf{q}_s)(T_s + 1) = 0, \quad$ if $s \in S$; $\qquad T_w T_u = T_{wu}, \quad$ if $l(wu) = l(w) + l(u)$.

The generic Hecke algebra of W actually can be defined over $\mathbb{Z}[\mathbf{q}_s]_{s \in S}$, but it is convenient to define it over $\mathcal{A}$ for introducing Kazhdan-Lusztig polynomials and for defining cells in W.

Let $\mathcal{H}'$ be the subalgebra of $\mathcal{H}$ generated by T_s, $s \in S$. Then the algebra $\mathcal{H}$ is isomorphic to the "twisted" tensor product $\mathbb{Z}[\Omega] \otimes_{\mathbb{Z}} \mathcal{H}'$ by assigning $T_{\omega w} \to \omega \otimes T_w$, where $\mathbb{Z}[\Omega]$ is the group algebra of Ω over $\mathbb{Z}$, and the multiplication in $\mathbb{Z}[\Omega] \otimes_{\mathbb{Z}} \mathcal{H}'$ is given by

$$(\omega \otimes T_w)(\omega' \otimes T_u) = \omega \omega' \otimes T_{\omega'^{-1}(w)} T_u.$$

Note that $s, t \in S$ may be conjugate in W but not conjugate in W', thus $\mathcal{H}'$ may not be the generic Hecke algebra of W' in the previous sense.

For an arbitrary $\mathcal{A}$-algebra $\mathcal{A}'$, the $\mathcal{A}'$-algebra $\mathcal{H} \otimes_{\mathcal{A}} \mathcal{A}'$ is called a Hecke algebra.

Convention: For each element w in W we shall denote the image in $\mathcal{H} \otimes_{\mathcal{A}} \mathcal{A}'$ of T_w by the same notation.

1.2. Two special choices of $\mathcal{A}'$ are of particular interests.

(a). Let $\mathbf{q}^{\frac{1}{2}}$ be an indeterminate and let $A = \mathbb{Z}[\mathbf{q}^{\frac{1}{2}}, \mathbf{q}^{-\frac{1}{2}}]$ be the ring of all Laurant polynomials in $\mathbf{q}^{\frac{1}{2}}$ with integer coefficients. Choose integers c_s, $s \in S$ such that $c_s = c_t$ whenever s and t are conjugate in W. There is a unique ring homomorphism from $\mathcal{A}$ to A such that $\mathbf{q}^{\frac{1}{2}}$ maps to $\mathbf{q}^{\frac{c_s}{2}}$ for every $s \in S$. Thus A is an $\mathcal{A}$-algebra. The multiplication laws in the Hecke algebra $\mathcal{H} \otimes_{\mathcal{A}} A$ are (recall the convention at the end of 1.1)

$$(1.2.1) \quad (T_s - \mathbf{q}^{c_s})(T_s + 1) = 0, \quad \text{if } s \in S; \qquad T_w T_u = T_{wu}, \quad \text{if } l(wu) = l(w) + l(u).$$

(b). When all integers c_s ($s \in S$) are 1, we denote the Hecke algebra $\mathcal{H} \otimes_{\mathcal{A}} A$ by H. The multiplication laws in H are

$$(1.2.2) \quad (T_s - \mathbf{q})(T_s + 1) = 0, \quad \text{if } s \in S; \qquad T_w T_u = T_{wu}, \quad \text{if } l(wu) = l(w) + l(u).$$

Sometimes H is also called the generic Hecke algebra of W (with one parameter). By now the Hecke algebra H and its various specializations $H \otimes_A A'$ are the most extensively studied Hecke algebras.

There is also a slight generalization of the Hecke algebra $\mathcal{H}$. Let R be a commutative ring with 1. For every s in S, choose u_s, $v_s \in R$ such that $u_s = u_t$, $v_s = v_t$ whenever s, t are conjugate in W. Then there exists a unique associative R-algebra $\tilde{\mathcal{H}}$, which is a free R-module with a basis T'_w, $w \in W$ and multiplication is given by

$$(1.2.3) \quad T'^2_s = u_s T'_s + v_s, \quad \text{if } s \in S; \qquad T'_w T'_u = T'_{wu}, \quad \text{if } l(wu) = l(w) + l(u).$$

(see, e.g. [Hu]). It is often that the R-algebra $\tilde{\mathcal{H}}$ is actually a Hecke algebra. Suppose that v_s has a square root $v_s^{\frac{1}{2}}$ in R and $v_s^{\frac{1}{2}}$ is invertible in R. Further we assume that there exists an invertible element $u_s'^{\frac{1}{2}} \in R$ such that

$$(1.2.4) \qquad\qquad u_s'^{\frac{1}{2}} - u_s'^{-\frac{1}{2}} = u_s v_s^{-\frac{1}{2}}.$$

Set $T''_s := u_s'^{\frac{1}{2}} v_s^{-\frac{1}{2}} T'_s$, then $T''^2_s = (u'_s - 1) T''_s + u'_s$. In this case the algebra $\tilde{\mathcal{H}}$ is a Hecke algebra in the sense of 1.1.

In Lie theory there are also other interesting algebras of Hecke type, see for example, [BM, Ca, MS].

1.3. Examples of Coxeter groups. It is convenient to represent a Coxeter system (W', S) by a graph Σ, usually called the Coxeter graph of (W', S). The vertex set of Σ is one to one correspondence with S; a pair of vertices corresponding to s_i, s_j are jointed with an edge whenever $m_{ij} \geq 3$, and label such an edge with m_{ij} when $m_{ij} \geq 4$. Thus the graph Σ determines (W', S) up to an isomorphism.

A Coxeter system (W', S) is called irreducible if for any $s, t \in S$ we can find a sequence $s = t_0, t_1, \cdots, t_k = t$ in S such that $m_{t_i, t_{i+1}} \geq 3$ (i.e. $t_i t_{i+1}$ is not equal to $t_{i+1} t_i$) for $i = 0, 1, ..., k - 1$. We also call W' an irreducible Coxeter group when (W', S) is irreducible. Obviously every Coxeter group is a direct product of some irreducible Coxeter groups.

The most important Coxeter groups in Lie theory are Weyl groups and affine Weyl groups. They are classified. The Coxeter graphs of irreducible Weyl groups and irreducible affine Weyl groups are as follows.

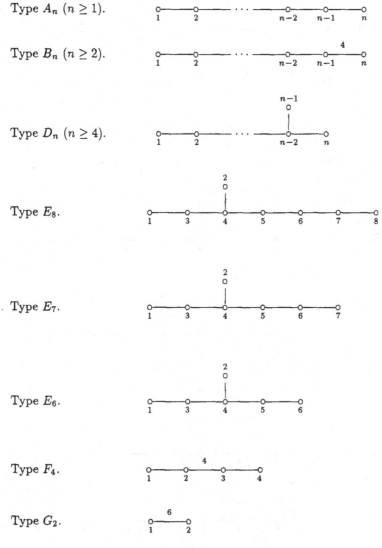

Type A_n $(n \geq 1)$.

Type B_n $(n \geq 2)$.

Type D_n $(n \geq 4)$.

Type E_8.

Type E_7.

Type E_6.

Type F_4.

Type G_2.

Type $\tilde{A}_1$.

Type $\tilde{A}_n$ $(n \geq 2)$.
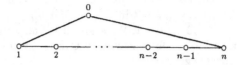

Type $\tilde{B}_2 = \tilde{C}_2$.

Type $\tilde{B}_n$ $(n \geq 3)$.

Type $\tilde{C}_n$ $(n \geq 3)$.

Type $\tilde{D}_n$ $(n \geq 4)$.
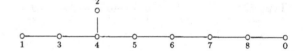

Type $\tilde{E}_8$.
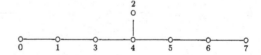

Type $\tilde{E}_7$.
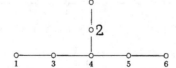

Type $\tilde{E}_6$.
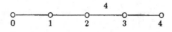

Type $\tilde{F}_4$.

4

Type $\tilde{G}_2$.

$$\underset{0}{\circ}\!\!-\!\!-\!\!-\!\!\underset{1}{\circ}\!\!\overset{6}{-\!\!-\!\!-}\!\!\underset{2}{\circ}$$

1.4. The Weyl group of type A_n is just the symmetric group $\mathfrak{S}_{n+1}$ of degree $n+1$. One may choose $\{(12), (23), \cdots, (n, n+1)\}$ as the set of simple reflections of $\mathfrak{S}_{n+1}$.

Except Weyl groups, the other irreducible finite Coxeter groups are dihedral groups $I_2(m)$ ($m = 5$ or $m > 6$, when $m = 3, 4, 6$, $I_2(m)$ are Weyl groups) and Coxeter groups of type H_3 or H_4. Their Coxeter graphs are as follows.

Type H_4.

$$\underset{1}{\circ}\!\!\overset{5}{-\!\!-\!\!-}\!\!\underset{2}{\circ}\!\!-\!\!-\!\!-\!\!\underset{3}{\circ}\!\!-\!\!-\!\!-\!\!\underset{4}{\circ}$$

Type H_3.

$$\underset{1}{\circ}\!\!\overset{5}{-\!\!-\!\!-}\!\!\underset{2}{\circ}\!\!-\!\!-\!\!-\!\!\underset{3}{\circ}$$

Type $I_2(m)$.

$$\underset{1}{\circ}\!\!\overset{m}{-\!\!-\!\!-}\!\!\underset{2}{\circ}$$

When (W', S) is crystallographic (i.e., $m_{ij} = 2, 3, 4, 6, \infty$ for arbitrary s_i, s_j in S), W' can be realized as the Weyl group of certain Kac-Moody algebra (see [K]). Thus we have a Schubert variety $\mathcal{B}_w$ for each element $w \in W'$. This is a key to apply the powerful intersection cohomology theory to the Kazhdan-Lusztig theory.

1.5. Examples of Hecke algebras. (a). Let G be a Chevalley group over a finite field of q elements. Let B be a Borel subgroup of G and T the maximal torus in B. Then the group $W_0 = N_G(T)/T$ is a Weyl group. We have $G = \bigcup_{w \in W_0} BwB$. Let H be the free $\mathbb{Z}$-module generated by the double cosets BwB, $w \in W_0$. We denote T_w the double coset BwB when it is regarded as an element in H. Define the multiplication in H by

$$(1.5.1) \qquad T_w T_u = \sum_v m_{w,u,v} T_v,$$

where the structure constants $m_{w,u,v}$ are defined as the number of cosets of the form Bx in the set $Bw^{-1}Bv \cap BuB$:

$$m_{w,u,v} = |Bw^{-1}Bv \cap BuB/B|.$$

Then H is an associative ring with unit T_e, where e is the neutral element in W_0. Moreover we have

$$(1.5.2) \ (T_s - q)(T_s + 1) = 0, \quad \text{if } s \in S_0; \qquad T_w T_u = T_{wu}, \quad \text{if } l(wu) = l(w) + l(u),$$

where S_0 is the set of simple reflections in W_0. (See [I]).

It is well known that $H \otimes_{\mathbb{Z}} \mathbb{C} \simeq \mathrm{End}1_B^G$, where 1_B^G stands for the induced representation of the unit representation 1_B (over $\mathbb{C}$) of B (see [I, C2, Cu]). Thus part of the study of 1_B^G can be reduced to that of $H \otimes_{\mathbb{Z}} \mathbb{C}$.

5

(b). Let K be a p-adic field such that its residue field k contains q elements. Let G be a Chevalley group over the field K. Let B be an Iwahori subgroup of G and T the 'diagonal' subgroup of B (in a suitable sense). Then the group $W = N_G(T)/T$ is an extended affine Weyl group (i.e., there is a commutative group Ω which acts on an affine Weyl group (W', S) such that $W \simeq \Omega \ltimes W'$, see 2.1 for definition). As the above example we have $G = \bigcup_{w \in W} BwB$. Let H be the free $\mathbb{Z}$-module generated by the double cosets BwB, $w \in W$. We denote T_w the double coset BwB when it is regarded as an element in H. Define the multiplication in H by

(1.5.3)
$$T_w T_u = \sum_v m_{w,u,v} T_v,$$

where the structure constants $m_{w,u,v}$ are defined as the number of cosets of the form Bx in the set $Bw^{-1}Bv \cap BuB$:

$$m_{w,u,v} = |Bw^{-1}Bv \cap BuB/B|.$$

Then H is an associative ring with unit T_e, where e is the neutral element in W. Moreover we have

(1.5.4) $(T_s - q)(T_s + 1) = 0$, if $s \in S$; $T_w T_u = T_{wu}$, if $l(wu) = l(w) + l(u)$,

where S is the set of simple reflections in W. (See [IM, p.44]).

It is known that the category of admissible complex representations of G which have nonzero vectors fixed by B is equivalent to the category of finite dimensional representations (over $\mathbb{C}$) of $H \otimes_{\mathbb{Z}} \mathbb{C}$ (see [Bo1, M]). Thus the representations of $H \otimes_{\mathbb{Z}} \mathbb{C}$ may be ragarded as an interesting part of the representation theory of p-adic groups.

1.6. Kazhdan-Lusztig polynomials. The work [KL1] stimulates a lot of work and deeply increased our understanding of Coxeter groups and of Hecke algebras. The key role is the Kazhdan-Lusztig polynomials. In this section we recall some definitions and results in [KL1].

We keep the notations in 1.1 and in 1.2 (b). Thus (W', S) is a Coxeter system, $W = \Omega \ltimes W'$ is an extended Coxeter group and H is the generic Hecke algebra of W over $A = \mathbb{Z}[\mathbf{q}^{\frac{1}{2}}, \mathbf{q}^{-\frac{1}{2}}]$.

Let $a \to \bar{a}$ be the involution of the ring A defined by $\bar{\mathbf{q}}^{\frac{1}{2}} = \mathbf{q}^{-\frac{1}{2}}$. This extends to an involution $h \to \bar{h}$ of the ring H defined by

$$\overline{\sum a_w T_w} = \sum \bar{a}_w T_{w^{-1}}^{-1}, \quad a_w \in A.$$

Note that T_w is invertible for any $w \in W$ since $T_s^{-1} = \mathbf{q}^{-1} T_s + (\mathbf{q}^{-1} - 1)$ for $s \in S$ and $T_\omega^{-1} = T_{\omega^{-1}}$ for $\omega \in \Omega$. Then (see [KL1, (1.1.c)]):

(a) For each $w \in W$, there is a unique element $C_w \in H$ such that

$$\bar{C}_w = C_w,$$

6

$$C_w = \mathbf{q}^{-\frac{l(w)}{2}} \sum_{y \le w} P_{y,w} T_y,$$

where $P_{y,w} \in A$ is a polynomial in $\mathbf{q}$ of degree $\le \frac{1}{2}(l(w) - l(y) - 1)$ for $y < w$ and $P_{w,w} = 1$.

The assertion (a) is equivalent to the following result.

(b) For each $w \in W$, there is a unique element $C'_w \in H$ such that $\bar{C}'_w = C'_w$ and $C'_w = \sum_{y \le w} (-1)^{l(w)-l(y)} \mathbf{q}^{\frac{l(w)}{2}} \mathbf{q}^{-l(y)} \bar{P}_{y,w} T_y$, where $P_{y,w} \in A$ is a polynomial in $\mathbf{q}$ of degree $\le \frac{1}{2}(l(w) - l(y) - 1)$ for $y < w$ and $P_{w,w} = 1$.

Note that our notations C_w and C'_w exchange these in [KL1] since we shall mainly use the elements C_w.

Obviously the elements C_w, $w \in W$ form an A-basis of H and the elements C'_w, $w \in W$ also form an A-basis of H. They are related by three involutions.

(c) Let j be the involution of the ring H given by

$$j(\sum a_w T_w) = \sum \bar{a}_w (-\mathbf{q})^{-l(w)} T_w,$$

then $C'_w = (-1)^{l(w)} j(C_w)$. (See [KL1]).

(d) Let Φ be the involution of the ring H defined by

$$\Phi(\mathbf{q}^{\frac{1}{2}}) = -\mathbf{q}^{\frac{1}{2}}, \ \Phi(T_w) = (-\mathbf{q})^{l(w)} T_{w^{-1}}^{-1},$$

then $C'_w = \Phi(C_w)$. (See [L11, 3.2, p.259]).

(e) Let k be the involution of the A-algebra H given by

$$k(\sum a_w T_w) = \sum a_w (-\mathbf{q})^{l(w)} T_{w^{-1}}^{-1},$$

then $C'_w = (-1)^{l(w)} k(C_w)$.

We give a proof of (e). It is easy to see that

$$k(\sum a_w T_w) = j(\overline{\sum a_w T_w}).$$

That is, k is the composition of j and $\bar{\cdot}$. Since $\bar{C}_w = C_w$, according to (c) we get $k(C_w) = j(\bar{C}_w) = j(C_w) = (-1)^{l(w)} C'_w$.

Note that k is an involution of A-algebra, this fact is useful in transferring some properties of C'_w to C_w.

The polynomials $P_{y,w}$ are called Kazhdan-Lusztig polynomials. For $y < w$ we have $P_{y,w} = \mu(y,w) \mathbf{q}^{\frac{1}{2}(l(w)-l(y)-1)}$+lower degree terms. We say that $y \prec w$ if $\mu(y,w) \ne 0$, we then set $\mu(w,y) = \mu(y,w)$.

7

1.7. Motivated by his definition of canonical bases of quantum groups (see [L19]), Lusztig gave another construction of the elements C_w, C'_w. Consider the $\mathbb{Z}[\mathbf{q}^{-\frac{1}{2}}]$-submodule $\mathcal{L}$ of H spanned by $\tilde{T}_w = \mathbf{q}^{-\frac{l(w)}{2}}T_w$, $w \in W$ and the $\mathbb{Z}[\mathbf{q}^{\frac{1}{2}}]$-submodule $\mathcal{L}'$ of H spanned by $\tilde{T}_w$, $w \in W$, then (see [L20])

(a) The projection $\pi : \mathcal{L} \to \mathcal{L}/\mathbf{q}^{-\frac{1}{2}}\mathcal{L}$ gives rise to an isomorphism of $\mathbb{Z}$-module $\pi_1 : \mathcal{L} \cap \bar{\mathcal{L}} \tilde{\to} \mathcal{L}/\mathbf{q}^{-\frac{1}{2}}\mathcal{L}$ and $\pi_1^{-1}(\pi(\tilde{T}_w)) = C_w$.

(b) The projection $\pi' : \mathcal{L}' \to \mathcal{L}'/\mathbf{q}^{\frac{1}{2}}\mathcal{L}'$ gives rise to an isomorphism of $\mathbb{Z}$-module $\pi_1 : \mathcal{L}' \cap \bar{\mathcal{L}}' \tilde{\to} \mathcal{L}'/\mathbf{q}^{\frac{1}{2}}\mathcal{L}'$ and $\pi_1'^{-1}(\pi'(\tilde{T}_w)) = C'_w$.

1.8. The elements C_w have the following properties (see [KL1]):

(a) For $s \in S$ we have

$$
C_s C_w = \begin{cases} (\mathbf{q}^{\frac{1}{2}} + \mathbf{q}^{-\frac{1}{2}})C_w, & \text{if } sw \leq w \\ C_{sw} + \sum_{\substack{y \prec w \\ sy \leq y}} \mu(y,w)C_y, & \text{if } sw \geq w. \end{cases}
$$

$$
C_w C_s = \begin{cases} (\mathbf{q}^{\frac{1}{2}} + \mathbf{q}^{-\frac{1}{2}})C_w, & \text{if } ws \leq w \\ C_{ws} + \sum_{\substack{y \prec w \\ ys \leq y}} \mu(y,w)C_y, & \text{if } ws \geq w. \end{cases}
$$

They are equivalent to the following recursion formulas of the Kazhdan-Lusztig polynomials.

(b) Assume that for $s,t \in S$ we have $sw > w$ and $wt > w$, then

$$
P_{y,sw} = \mathbf{q}^{1-a}P_{sy,w} + \mathbf{q}^a P_{y,w} - \sum_{\substack{z \\ y \leq z \prec w \\ sz < z}} \mu(z,w)\mathbf{q}^{\frac{l(w)-l(z)+1}{2}}P_{y,z}, \quad (y \leq sw)
$$

where $a = 1$ if $sy < y$, $a = 0$ if $sy > y$; and $P_{y,sw} = P_{sy,sw}$.

$$
P_{y,wt} = \mathbf{q}^{1-a}P_{yt,w} + \mathbf{q}^a P_{y,w} - \sum_{\substack{z \\ y \leq z \prec w \\ zt < z}} \mu(z,w)\mathbf{q}^{\frac{l(w)-l(z)+1}{2}}P_{y,z}, \quad (y \leq wt)
$$

where $a = 1$ if $yt < y$, $a = 0$ if $yt > y$; and $P_{y,wt} = P_{yt,wt}$.

1.9. When (W',S) is a finite Coxeter group or a crystallographic group, it is known that the coefficients of $P_{y,w}$ are non-negative. This is proved in [KL2, L11] when (W',S) is crystallographic. For H_3, H_4 it was proved by Goresky [Go] and Alvis [A]. For dihedral groups I_m it is trivial since $P_{y,w} = 1$ for any $y \leq w$. It was conjectured in [KL1] that for an arbitrary Coxeter group the Kazhdan-Lusztig polynomials have non-negative coefficients.

1.10. Question. (i). *It is known that the Kazhdan-Lusztig polynomials of crystallographic Coxeter groups are related to middle intersection cohomology groups*

8

of Schubert varieties. Now what polynomials are related to other intersection cohomology groups of Schubert varieties?

(ii). If we loose the restriction on the degree of $P_{y,w}$ to $\deg P_{y,w} \leq (l(w) - l(y))$, what happen for the Kazhdan-Lusztig polynomials and the elements C_w.

1.11. Cell   For each element w in W we set

$$L(w) := \{s \in S \mid sw < w\},$$

$$R(w) := \{s \in S \mid ws < w\}.$$

Let w and u be elements in W', we say that $w \underset{L}{\leq} u$ (resp. $w \underset{R}{\leq} u$; $w \underset{LR}{\leq} u$) if there exists a sequence $w = w_0, w_1, ..., w_k = u$ in W' such that for $i = 1, 2, ...k$ we have $\mu(w_{i-1}, w_i) \neq 0$ and $L(w_{i-1}) \not\subseteq L(w_i)$ (resp. $R(w_{i-1}) \not\subseteq R(w_i)$; $L(w_{i-1}) \not\subseteq L(w_i)$ or $R(w_{i-1}) \not\subseteq R(w_i)$). Then for any $\omega, \omega' \in \Omega$ we say that $\omega w \underset{L}{\leq} \omega' u$ (resp. $w\omega \underset{R}{\leq} u\omega'$; $\omega w \underset{LR}{\leq} \omega' u$) if $w \underset{L}{\leq} u$ (resp. $w \underset{R}{\leq} u$; $w \underset{LR}{\leq} u$).

For elements x and y in W we write that $x \underset{L}{\sim} y$ (resp. $x \underset{R}{\sim} y$; $x \underset{LR}{\sim} y$) if $x \underset{L}{\leq} y \underset{L}{\leq} x$ (resp. $x \underset{R}{\leq} y \underset{R}{\leq} x$; $x \underset{LR}{\leq} y \underset{LR}{\leq} x$). The relations $\underset{L}{\leq}$, $\underset{R}{\leq}$, $\underset{LR}{\leq}$ are preorders in W. And the relations $\underset{L}{\sim}$, $\underset{R}{\sim}$, $\underset{LR}{\sim}$ are equivalence relations in W, the corresponding equivalence classes are called left cells, right cells, two-sided cells of W, respectively. The preorder $\underset{L}{\leq}$ (resp. $\underset{R}{\leq}$; $\underset{LR}{\leq}$) induces a partial order on the set of left (resp. right; two-sided) cells of W, we denote it again by $\underset{L}{\leq}$ (resp. $\underset{R}{\leq}$; $\underset{LR}{\leq}$).

When $W = W'$ is a Weyl group, the definitions of left cell and two-sided cell coincide with the definitions given by Joseph [J1-J2]. The cells in Weyl groups were extensively investigated by Barbasch, Lusztig, Joseph, Vogan, etc., and play an important role in the representation theory of finite groups of Lie type (see [L7]) and in the theory of primitive ideals of universal enveloping algebras of semisimple Lie algebras.

For affine Weyl groups, the structure of left cells and two-sided cells are determined for type $\tilde{A}_n$ (see [Sh1, L8]), rank 2, 3 (see [L11, Bé1, D]). Recently Shi found an algorithm, then he and his students determined the structure of cells in affine Weyl groups of type $\tilde{B}_4$, $\tilde{C}_4$, $\tilde{D}_4$ (see [Sh4, Sh5]). For type $\tilde{D}_4$, see also [Ch]. In [L11-L14] Lusztig obtained a series of important results concerned with cells in affine Weyl groups.

1.12. a-function   For an extended Coxeter group W the function $a : W \to \mathbb{N}$ was introduced in [L11] and is a useful tool in cell theory and related topics.

Given $w, u \in W$, we write

$$C_w C_u = \sum_{v \in W} h_{w,u,v} C_v, \qquad h_{w,u,v} \in A.$$

9

For every $v \in W$, we define $a(v)=$the minimal non-negative integer i such that $\mathbf{q}^{\frac{i}{2}} h_{w,u,v}$ is in $\mathbb{Z}[\mathbf{q}^{\frac{1}{2}}]$ for any $w, u \in W$. If such i doesnot exist, we set $a(v) = \infty$.

For a finite Coxeter group, the function a is always bounded. A non-trivial fact is that a is bounded for an affine Weyl group (see [L11]). In [L12] Lusztig obtained some interesting results provided that a is bounded and W' is crystallographic.

Assume that (W', S) is a crystallographic group, then all $h_{w,u,v}$ are Laurant polynomials in $\mathbf{q}^{\frac{1}{2}}$ with the same purity and have non-negative coefficients (see [L11]). It seems natural to hope such property holds for arbitrary Coxeter groups.

Here are four questions.

1.13. Question. (i). *Find out all Coxeter groups whose a-functions are bounded.*

(ii). *Assume that the a-function of a Coxeter group W is bounded and let a_0 be the maximal value of a on W. Is the set $\{w \in W \mid a(w) = a_0\}$ a two-sided cell of W?*

(iii). *Find out a Coxeter group W' such that there exists some $w \in W'$ with $a(w) = \infty$.*

(iv). *Maybe the a-function of a Coxeter group W is always bounded and the maximal value is equal to the length of the longest elements of certain finite Coxeter (or parabolic) subgroups of W.*

Generalized Cells

1.14. Lusztig generalized the definition of cells in [KL1] to the cases of simple reflections being given different weights (see [L6]). Strangely the interesting generalization is less developed. In the rest of the chapter we shall give some discussions to the generalization. We first recall the definition, then show how to apply the definition.

Let (W', S) be a Coxeter system and $W = \Omega \ltimes W'$ be an extended Coxeter group. Let $\varphi: W \to \Gamma$ be a map from W into an abelian group Γ such that $\varphi(\omega s_1 s_2 \cdots s_k \omega') = \varphi(s_1)\varphi(s_2)\cdots\varphi(s_k)$ for any reduced expression $s_1 s_2 \cdots s_k$ in W' and ω, ω' in Ω. Note that $\varphi(w) = \varphi(w')$ whenever w, w' are conjugate in W. For each w in W we shall write $\mathbf{q}_w^{\frac{1}{2}}$ for $\varphi(w)$. Let H_φ be the Hecke algebra of W with respect to φ; this is an associative algebra over the group ring $\mathbb{Z}[\Gamma]$. As a $\mathbb{Z}[\Gamma]$-module, it is free with a basis T_w, $w \in W$. The multiplication is defined by

$$(1.14.1) \quad (T_s - \mathbf{q}_s)(T_s + 1) = 0, \quad \text{if } s \in S; \qquad T_w T_u = T_{wu}, \quad \text{if } l(wu) = l(w) + l(u).$$

When $\mathbf{q}_s^{\frac{1}{2}} = \mathbf{q}_t^{\frac{1}{2}}$ if and only if s, t are conjugate in W and Γ is a free abelian group with a basis $\mathbf{q}_s^{\frac{1}{2}}$, $s \in S$, the algebra H_φ is canonically isomorphic to the algebra $\mathcal{H}$ in 1.1 if we identify $\mathbb{Z}[\Gamma]$ with $\mathcal{A}$. When $\mathbf{q}_s^{\frac{1}{2}} = \mathbf{q}_t^{\frac{1}{2}}$ for any $s, t \in S$, and Γ is a free abelian group generated by $\mathbf{q}_s^{\frac{1}{2}}$, the algebra H_φ is canonically isomorphic to the algebra H in 1.2 (b). Suitably choose the map $\varphi : W \to < \mathbf{q}_s^{\frac{1}{2}} >$ (the free abelian group generated by $\mathbf{q}_s^{\frac{1}{2}}$) we see that the Hecke algebra in 1.2 (a) is also canonically isomorphic to some H_φ.

It will be convenient to introduce a new basis $\tilde{T}_w = \mathbf{q}_w^{-\frac{1}{2}} T_w$ ($w \in W$) of H_φ. We then have

$$(\tilde{T}_s + \mathbf{q}_s^{-\frac{1}{2}})(\tilde{T}_s - \mathbf{q}_s^{\frac{1}{2}}) = 0, \qquad (s \in S).$$

Let $a \to \bar{a}$ be the involution of the ring $\mathbb{Z}[\Gamma]$ which takes γ to γ^{-1} for any $\gamma \in \Gamma$. This extends to an involution $h \to \bar{h}$ of the ring H_φ defined by

$$\overline{\sum a_w \tilde{T}_w} = \sum \bar{a}_w \tilde{T}_{w^{-1}}^{-1} \qquad (a_w \in \mathbb{Z}[\Gamma]).$$

Note that $\tilde{T}_w$ is invertible for any $w \in W$ since $\tilde{T}_s^{-1} = \tilde{T}_s + (\mathbf{q}_s^{-\frac{1}{2}} - \mathbf{q}_s^{\frac{1}{2}})$ for $s \in S$ and $\tilde{T}_\omega^{-1} = \tilde{T}_{\omega^{-1}}$ for $\omega \in \Omega$. We define the elements $R_{x,y}^* \in \mathbb{Z}[\Gamma]$, $(x, y \in W)$, by

$$\tilde{T}_{y^{-1}}^{-1} = \sum_x \bar{R}_{x,y}^* \tilde{T}_x.$$

It is easy to see that $R_{x,y}^* = 0$ unless $x \leq y$ in the standard partial order of W. Using the fact that $h \to \bar{h}$ is an involution, we see that

(1.14.2) $$\sum_{x \leq y \leq z} \bar{R}_{x,y}^* R_{y,z}^* = \delta_{x,z},$$

for all $x \leq z$ in W. Note also that $\mathbf{q}_x^{-\frac{1}{2}} \mathbf{q}_y^{\frac{1}{2}} R_{x,y}^* \in \mathbb{Z}[\Gamma^2]$ (convention: $\Gamma^2 = \{\gamma^2 \mid \gamma \in \Gamma\}$).

(1.14.3). $R_{x,x}^* = 1$ for all x in W.

(1.14.4). If $x < y$ and $l(y) = l(x) + 1$, then x is obtained by dropping some $s \in S$ in a reduced expression of y, and we have $R_{x,y}^* = \mathbf{q}_s^{\frac{1}{2}} - \mathbf{q}_s^{-\frac{1}{2}}$.

(1.14.5). If $x < y$ and $l(y) = l(x) + 2$, then x is obtained by dropping some $s, t \in S$ in a reduced expression of y, and we have $R_{x,y}^* = (\mathbf{q}_s^{\frac{1}{2}} - \mathbf{q}_s^{-\frac{1}{2}})(\mathbf{q}_t^{\frac{1}{2}} - \mathbf{q}_t^{-\frac{1}{2}})$.

We now assume that a total order Σ on Γ is given which is compatible with the group structure on Γ. Let Γ_+ be the set of elements which are strictly positive (i.e. bigger than the neutral element) for this total order and let $\Gamma_- = (\Gamma_+)^{-1}$. We shall assume that $\mathbf{q}_s^{\frac{1}{2}} \in \Gamma_+$ for all simple reflections s in S and *then the pair (Σ, φ) will be called admissible*. We have (see 2. Proposition in [L6])

(a) For each element w in W, there is a unique element $C_w \in H_\varphi$ such that

$$\bar{C}_w = C_w,$$

$$C_w = \sum_{y \leq w} P_{y,w}^* \tilde{T}_y,$$

11

where $P^*_{y,w} \in \mathbb{Z}[\Gamma]$ is a $\mathbb{Z}$-linear combination of elements in Γ_- for $y < w$ and $P^*_{w,w} = 1$. Moreover $\mathbf{q}_y^{-\frac{1}{2}} \mathbf{q}_w^{\frac{1}{2}} P^*_{y,w} \in \mathbb{Z}[\Gamma^2]$.

(When φ is constant on S, this is the same as (1.1.c) of [KL1].)

We also can introduce the elements C'_w as 1.6(b) and show that C_w, C'_w are related by three involutions of H_φ as (c-e) in 1.6.

Now let $s \in S$ and $w \in W$ be such that $w < sw$. For each y such that $sy < y < w$, we define an element

$$M^s_{y,w} \in \mathbb{Z}[\Gamma]$$

by the inductive condition

$$(1.14.6) \qquad \sum_{\substack{y \leq z < w \\ sz < z}} P^*_{y,z} M^s_{z,w} - \mathbf{q}_s^{\frac{1}{2}} P^*_{y,w} \quad \text{is a combination of elements in } \Gamma_-$$

and by the symmetry condition

$$(1.14.7) \qquad \bar{M}^s_{y,w} = M^s_{y,w}.$$

The condition (1.14.6) determines uniquely the coefficient of γ in $M^s_{y,w}$ for every γ in $\Gamma - \Gamma_-$; the condition (1.14.7) determines other coefficients. We have

$$\mathbf{q}_s^{-\frac{1}{2}} \mathbf{q}_y^{-\frac{1}{2}} \mathbf{q}_w^{\frac{1}{2}} M^s_{y,w} \in \mathbb{Z}[\Gamma^2].$$

Let $s \in S$ and $w \in W$, then (see 4. Proposition in [L6])

$$(1.14.8) \qquad (\tilde{T}_s + \mathbf{q}_s^{-\frac{1}{2}})C_w = C_{sw} + \sum_{\substack{z < w \\ sz < z}} M^s_{z,w} C_z, \qquad \text{if } w < sw;$$

$$(1.14.9) \qquad (\tilde{T}_s - \mathbf{q}_s^{\frac{1}{2}})C_w = 0, \qquad \text{if } w > sw.$$

Let j' be the anti-automorphism of the ring H_φ defined by $j'(\tilde{T}_w) = \tilde{T}_{w^{-1}}$ and $j'(a) = a$ for $a \in \mathbb{Z}[\Gamma]$. It is easy to see that $j'(C_w) = C_{w^{-1}}$. From (1.14.8-9) we can deduce

$$(1.14.10) \qquad C_w(\tilde{T}_s + \mathbf{q}_s^{-\frac{1}{2}}) = C_{ws} + \sum_{\substack{z < w \\ zs < z}} M^s_{z^{-1},w^{-1}} C_z, \qquad \text{if } w < ws;$$

$$(1.14.11) \qquad C_w(\tilde{T}_s - \mathbf{q}_s^{\frac{1}{2}}) = 0, \qquad \text{if } w > ws.$$

The identity (1.14.9) is equivalent to the following

(1.14.12) $\qquad P^*_{u,w} = \mathbf{q}_s^{-\frac{1}{2}} P^*_{su,w}$ $\qquad$ if $u < su \le w$, $sw < w$.

And the identity (1.14.11) is equivalent to the following

(1.14.13) $\qquad P^*_{u,w} = \mathbf{q}_s^{-\frac{1}{2}} P^*_{us,w}$ $\qquad$ if $u < us \le w$, $ws < w$.

(b) Let $y < w$ be such that $l(w) = l(y) + 1$. Then y is obtained by dropping a simple reflection s in a reduced expression of w. We have

(i). $P^*_{y,w} = \mathbf{q}_s^{-\frac{1}{2}}$.

(ii). Let t be a simple reflection such that $ty < y < w < tw$, then

$$M^t_{y,w} = \begin{cases} 0, & \text{if } \mathbf{q}_t^{\frac{1}{2}} < \mathbf{q}_s^{\frac{1}{2}}, \\ 1, & \text{if } \mathbf{q}_t^{\frac{1}{2}} = \mathbf{q}_s^{\frac{1}{2}}, \\ \mathbf{q}_s^{\frac{1}{2}}\mathbf{q}_t^{-\frac{1}{2}} + \mathbf{q}_s^{-\frac{1}{2}}\mathbf{q}_t^{\frac{1}{2}}, & \text{if } \mathbf{q}_t^{\frac{1}{2}} > \mathbf{q}_s^{\frac{1}{2}}. \end{cases}$$

1.15. Generalized cells. Let $w \in W$ be such that $w < sw$ for some $s \in S$. We shall write $z \underset{L,\varphi}{\le} w$ if $z = \omega w$ ($\omega \in \Omega$) or $z = sw$ or $M^s_{z,w} \ne 0$ (see (1.14.8)). We again use $\underset{L,\varphi}{\le}$ for the preorder relation on W generated by the relation $z \underset{L,\varphi}{\le} w$. The equivalence relation associated to the preorder $\underset{L,\varphi}{\le}$ is denoted by $\underset{L,\varphi}{\sim}$ and the corresponding equivalence classes in W are called generalized left cells of W (with respect to the pair (Σ, φ)). Given w, u in W, we say that $w \underset{LR,\varphi}{\le} u$ if there is a sequence $w = w_0, w_1, ..., w_k = u$ in W such that for $i = 0, 1, ..., k - 1$, we have either $w_i \underset{L,\varphi}{\le} w_{i+1}$ or $w_i^{-1} \underset{L,\varphi}{\le} w_{i+1}^{-1}$. The equivalence relation associated to the preorder $\underset{LR,\varphi}{\le}$ is denoted by $\underset{LR,\varphi}{\sim}$ and the corresponding equivalence classes in W are called generalized two-sided cells of W (with respect to the pair (Σ, φ)). (The notations $\underset{L,\varphi}{\le}, \underset{L,\varphi}{\sim}$, etc. should be replaced by $\underset{L,\Sigma,\varphi}{\le}, \underset{L,\Sigma,\varphi}{\sim}$, etc., but the later looks too complicated.) From the definitions and (1.14.8-11) we get

(a) Let $h, h' \in H_\varphi$ and $w \in W$. We write

$$hC_w = \sum_{u \in W} a_u C_u, \qquad a_u \in \mathbb{Z}[\Gamma],$$

$$hC_w h' = \sum_{u \in W} b_u C_u, \qquad b_u \in \mathbb{Z}[\Gamma].$$

Then $u \underset{L,\varphi}{\le} w$ if $a_u \ne 0$, and $u \underset{LR,\varphi}{\le} w$ if $b_u \ne 0$.

13

For every element w in W, we denote I_w^L (resp. $\hat{I}_w^L$) the $\mathbb{Z}[\Gamma]$-submodule of H_φ spanned by the elements C_u, $u \underset{L,\varphi}{\leq} w$, (resp. by the elements C_u, $u \underset{L,\varphi}{\leq} w$, $u \underset{L,\varphi}{\not\sim} w$). We denote similarly I_w^{LR} (resp. $\hat{I}_w^{LR}$) the $\mathbb{Z}[\Gamma]$-submodule of H_φ spanned by the elements C_u, $u \underset{LR,\varphi}{\leq} w$, (resp. by the elements C_u, $u \underset{LR,\varphi}{\leq} w$, $u \underset{LR,\varphi}{\not\sim} w$).

It is clear from (a) that I_w^L, $\hat{I}_w^L$ are left ideals of H_φ and I_w^{LR}, $\hat{I}_w^{LR}$ are two-sided ideals of H_φ. Hence $I_w^L/\hat{I}_w^L$ is a left H_φ-module with a natural basis given by the images of C_u for u in the generalized left cell containing w; $I_w^{LR}/\hat{I}_w^{LR}$ is a two-sided H_φ-module with a natural basis given by the images of C_u for u in the generalized two-sided cell containing w.

1.16. Assume that Γ is a free abelian group with a basis $\mathbf{q}_s^{\frac{1}{2}}$, $s \in S$ (note that $\mathbf{q}_s^{\frac{1}{2}} = \mathbf{q}_t^{\frac{1}{2}}$ whenever $s,t \in S$ are conjugate in W). It is known that we can totally order the set $Q_S = \{\mathbf{q}_s^{\frac{1}{2}} \mid s \in S\}$. Assume that on Q_S such a total order $\leq$ is given and at the moment we write $t < s$ if $\mathbf{q}_t^{\frac{1}{2}} < \mathbf{q}_s^{\frac{1}{2}}$. We define

$$\mathbf{q}_s^{-\frac{1}{2}} < \prod_{\substack{t<s \\ a_t \in \mathbb{Z}}} \mathbf{q}_t^{\frac{a_t}{2}} < \mathbf{q}_s^{\frac{1}{2}},$$

where only finitely many a_t $(t \in S)$ are nonzero, and define $\mathbf{q}_s^{\frac{i}{2}} < \mathbf{q}_s^{\frac{j}{2}}$ if and only if $i < j$. It is easy to see that we have defined a total order on the group Γ which is compatible with the group structure of Γ. Thus the definitions and results in 1.14-15 can be applied to the Hecke algebra H_φ.

If $\operatorname{Card}\{\mathbf{q}_s^{\frac{1}{2}} \mid s \in S\} \leq \operatorname{Card}\mathbb{R}$ (here $\mathbb{R}$ denotes the field of real numbers and Card denotes the cardinal of a set), we can find more natural total orders on Γ which are compatible with the group structure. Actually we can find many injective homomorphisms of abelian groups $\tau : \Gamma \to \mathbb{R}$ such that $\tau(\mathbf{q}_s^{\frac{1}{2}}) > 0$ for all s in S. The total order in $\tau(\Gamma)$ gives rise to a total order on Γ which is compatible with the group structure on Γ.

1.17. Proposition. *Assume that W is a finite Coxeter group and let S be the set of simple reflections in W. Let w_0 be the longest element in W. We have*

(i). *Let $h \in H_\varphi$ be such that $\tilde{T}_s h = \mathbf{q}_s^{\frac{1}{2}} h$ for any $s \in S$, then $h = a\sum_{w \in W} T_w$ for some $a \in \mathbb{Z}[\Gamma]$.*

(ii). $C_{w_0} = \mathbf{q}_{w_0}^{-\frac{1}{2}} \sum_{w \in W} T_w$.

(iii). *The set $\{w_0\}$ is a generalized two-sided cell of W (with respect to any admissible pair (Σ, φ)).*

(iv). *The set $\{e\}$ is a generalized two-sided cell of W (with respect to any admissible pair (Σ, φ)).*

Proof. (i). First we have $\tilde{T}_s \sum_{w \in W} T_w = \mathbf{q}_s^{\frac{1}{2}} \sum_{w \in W} T_w$. Let

$$h = a \sum_{w \in W} T_w + \sum_{\substack{w \in W \\ w < w_0}} a_w T_w, \qquad a, a_w \in \mathbb{Z}[\Gamma]$$

14

Assume that $a_w \neq 0$ for some $w < w_0$. Choose one such w with maximal length. Since $w < w_0$, we can find some $s \in S$ such that $w < sw$. Thus in the expression

$$\tilde{T}_s h = \mathbf{q}_s^{\frac{1}{2}} a \sum_{w \in W} T_w + \sum_{w \in W} b_w T_w, \qquad b_w \in \mathbb{Z}[\Gamma]$$

We have $b_{sw} = a_w \neq 0$, it contradicts to our assumptions on h and on w. Hence all $a_w = 0$ whenever $w < w_0$.

(ii). By (1.14.9) and (i) we see that $C_{w_0} = a \sum_{w \in W} T_w$ for some $a \in \mathbb{Z}[\Gamma]$. Note that $R^*_{x,y} = 0$ unless $x \leq y$ and that $R^*_{y,y} = 1$. From $\bar{C}_{w_0} = C_{w_0}$ we get $a = \mathbf{q}_{w_0}^{-\frac{1}{2}}$. The assertion is proved. One also can use (1.14.12) to prove (ii).

(iii). For any s in S we have $\tilde{T}_s C_{w_0} = C_{w_0} \tilde{T}_s = \mathbf{q}_s^{\frac{1}{2}} C_{w_0}$. By the definition of generalized two-sided cell we get the assertion.

(iv). It follows from (1.14.8), (1.14.10) and the definition of generalized two-sided cell.

1.18. Proposition. *Keep the set up in 1.14. Let s be a simple reflection in S and w an element in W. Then*

(i). $ws \leq w$ *if and only if* $C_w = h C_s$ *for some h in H_φ.*

(ii). $sw \leq w$ *if and only if* $C_w = C_s h$ *for some h in H_φ.*

Proof. (i). We use induction on $l(w)$. When $l(w) = 1$, the assertion is obvious. Now assume that $l(w) > 1$ and $ws \leq w$. According to (1.14.10),

$$C_{ws} C_s = C_w + \sum_{\substack{z < w \\ zs < z}} M^s_{z^{-1}, w^{-1}} C_z.$$

Using induction hypothesis we know $C_w = h C_s$ for some h in H_φ. The "only if" part is proved.

Suppose $C_w = h C_s$ for some h in H_φ. Using (1.14.10-11) and noting that $C_s \tilde{T}_s = \mathbf{q}_s^{\frac{1}{2}} C_s$, we see $ws \leq w$.

The proof of (ii) is similar.

1.19. Corollary. *Keep the set up in 1.14. Let S_1 be a subset of S and assume that S_1 generates a finite subgroup W_1 of W. Let w_1 be the longest element of W_1. Then the elements C_y, $y \in W$ and $l(y) = l(y w_1) + l(w_1)$, form a $\mathbb{Z}[\Gamma]$-basis of the left ideal $H_\varphi C_{w_1}$ of H_φ generated by C_{w_1}.*

1.20. Corollary. *Keep the set up in 1.14. Let $y, w \in W$, Assume that $y \underset{L,\varphi}{\leq} w$, then $R(w) \subseteq R(y)$. In particular $R(w) = R(y)$ if $y \underset{L,\varphi}{\sim} w$. (See 1.11 for the definition of $R(\ )$.)*

Proof. Let s be a simple reflection such that $ws \leq w$. By 1.18 (i), $C_w = h C_s$ for some h in H_φ. According to the definition of the preorder $\underset{L,\varphi}{\leq}$ and as the proof of 1.17 (i), we know $ys \leq y$. The corollary is proved.

15

1.21. Examples. Let (W, S) be a dihedral group of type $I_2(2m)$ (see 1.4). Let $s, t \in S$, then $(st)^{2m} = 1$. Let $\varphi : W \to \Gamma$ and H_φ be as in 1.14. We assume that $\mathbf{q}_t^{\frac{1}{2}} > \mathbf{q}_s^{\frac{1}{2}}$. We shall write $\mathbf{u}$, $\mathbf{v}$ instead of $\mathbf{q}_s^{\frac{1}{2}}$, $\mathbf{q}_t^{\frac{1}{2}}$ respectively.

(a). Let y, w be elements in W such that $sy < y < w < sw$, then $M_{y,w}^s = 0$.

Proof. There exist $w_1, y_1 \in W$ such that $w = t \cdot w_1$, $y = s \cdot y_1$. (Convention: $x = x_1 \cdot x_2$ means that $x = x_1 x_2$ and $l(x) = l(x_1) + l(x_2)$.) Using (1.14.12) we get $P_{y,w}^* = \mathbf{v}^{-1} P_{ty,w}^*$. Note that $\mathbf{v} > \mathbf{u}$, we see that $\mathbf{u} P_{y,w}^* = \mathbf{u}\mathbf{v}^{-1} P_{ty,w}^*$ is a $\mathbb{Z}$-combination of elements in Γ_-. Using induction on $l(w) - l(y)$ and using the definition of $M_{y,w}^s$ we obtain (a).

Consider the simplest example (see [L6]): type B_2. Then $(st)^4 = 1$. We have

$$P_{t,tst}^* = \mathbf{v}^{-1}(\mathbf{u}^{-1} - \mathbf{u}), \qquad P_{e,tst}^* = \mathbf{v}^{-2}(\mathbf{u}^{-1} - \mathbf{u}),$$

$$P_{s,sts}^* = \mathbf{v}^{-1}(\mathbf{u}^{-1} + \mathbf{u}), \qquad P_{e,sts}^* = \mathbf{u}^{-1}\mathbf{v}^{-1}(\mathbf{u}^{-1} + \mathbf{u}),$$

and $P_{y,w}^* = \mathbf{q}_y^{\frac{1}{2}} \mathbf{q}_w^{-\frac{1}{2}}$ for all other pairs $y \le w$. (In particular, $P_{y,w}^*$ may have negative coefficients.) We have

$$M_{ts,sts}^t = M_{t,st}^t = \mathbf{u}\mathbf{v}^{-1} + \mathbf{u}^{-1}\mathbf{v}.$$

The generalized left cells are $\{e\}$, $\{s\}$, $\{t, st\}, \{tst\}$, $\{ts, sts\}$, $\{stst\}$. The corresponding H_φ-modules $I_w^L / \hat{I}_w^L$ (with scalars extended to an algebraic closure of $\mathbf{Q}(\mathbf{u}, \mathbf{v})$) are all irreducible. (This is in contrast with the situation in [KL1].) The generalized two-sided cells are $\{e\}$, $\{s\}$, $\{tst\}$, $\{t, st, ts, sts\}$, $\{stst\}$.

The second simplest example is type G_2. Then $(st)^6 = 1$. We have

$$P_{t,tst}^* = P_{tst,tstst}^* = \mathbf{v}^{-1}(\mathbf{u}^{-1} - \mathbf{u}),$$

$$P_{e,tst}^* = P_{ts,tstst}^* = P_{st,tstst}^* = \mathbf{v}^{-2}(\mathbf{u}^{-1} - \mathbf{u}),$$

$$P_{t,tstst}^* = \mathbf{v}^{-2}(\mathbf{u}^{-2} - 1 + \mathbf{u}^2),$$

$$P_{e,tstst}^* = \mathbf{v}^{-3}(\mathbf{u}^{-2} - 1 + \mathbf{u}^2),$$

$$P_{s,sts}^* = P_{sts,ststs}^* = \mathbf{v}^{-1}(\mathbf{u}^{-1} + \mathbf{u}),$$

$$P_{e,sts}^* = P_{st,ststs}^* = P_{ts,ststs}^* = \mathbf{u}^{-1}\mathbf{v}^{-1}(\mathbf{u}^{-1} + \mathbf{u}),$$

$$P_{t,ststs}^* = \mathbf{u}^{-2}\mathbf{v}^{-1}(\mathbf{u}^{-1} + \mathbf{u}),$$

$$P_{s,ststs}^* = \mathbf{u}^{-1}\mathbf{v}^{-2}(\mathbf{u}^{-1} + \mathbf{u}),$$

$$P_{e,ststs}^* = \mathbf{u}^{-2}\mathbf{v}^{-2}(\mathbf{u}^{-1} + \mathbf{u}),$$

and $P_{y,w}^* = \mathbf{q}_y^{\frac{1}{2}} \mathbf{q}_w^{-\frac{1}{2}}$ for all other pairs $y \le w$. We have

$$M_{tsts,ststs}^t = M_{tst,stst}^t = M_{ts,sts}^t = M_{t,st}^t = \mathbf{u}\mathbf{v}^{-1} + \mathbf{u}^{-1}\mathbf{v},$$

$$M_{t,stst}^t = M_{ts,ststs}^t = 1.$$

16

The generalized left cells are

$$\{e\}, \{s\}, \{tstst\}, \{ststst\}, \{t, st, tst, stst\}, \{ts, sts, tsts, ststs\}.$$

The first four corresponding H_φ-modules $I_w^L / \hat{I}_w^L$ (with scalars extended to an algebraic closure of $\mathbf{Q}(\mathbf{u}, \mathbf{v})$) are all irreducible; the last two corresponding H_φ-modules are not irreducible. (This is in contrast with the situation in [KL1], also in contrast with the case type B_2.) The generalized two-sided cells are

$$\{e\}, \{s\}, \{t, st, ts, tst, sts, tsts, tsts, ststs\}, \{tstst\}, \{ststst\}.$$

1.22. Let (W, S) be a finite Coxeter group. When W is of type B_n, the generalized left cells of W provide all irreducible modules of the group algebra $\mathbb{C}[W]$. In general the generalized left cells of W should provide more irreducible modules of $\mathbb{C}[W]$ than the (ordinary) left cells of W. The work [L11-L14] showed that the cells in affine Weyl groups are interesting to the representations of affine Hecke algebras of one parameter. The generalized cells in affine Weyl groups should be interesting to the representations of affine Hecke algebras with unequal parameters.

It is not true that any generalized left (resp. two-sided) cell of an extended Coxeter group is always contained in a left (resp. two-sided) cell of the sense 1.11, as observed by H. Rui (a student of J. Shi). The following example is due to Rui.

Let (W, S) be a Weyl group of type B_3. We number the simple reflections in S so that $s_1 s_3 = s_3 s_1$, $(s_1 s_2)^3 = e$ and $(s_2 s_3)^4 = e$. Let $\varphi : W \to < \mathbf{q}^{\frac{1}{2}} >$ be such that $\varphi(wu) = \varphi(w)\varphi(u)$ if $l(wu) = l(w) + l(u)$, $\varphi(s_1) = \varphi(s_2) = \mathbf{q}^{\frac{1}{2}}$ and $\varphi(s_3) = \mathbf{q}^{\frac{3}{2}}$. We assume that $\mathbf{q}^{\frac{i}{2}} < \mathbf{q}^{\frac{j}{2}}$ if and only if $i < j$. Consider the elements $w = s_1 s_2 s_1 s_3$ and $y = s_1 s_3$. Obviously we have $w \underset{L,\varphi}{\leq} s_2 s_1 s_3 \underset{L,\varphi}{\leq} y$. A simple computation shows that $M_{y,w}^{s_3} = \mathbf{q}^{\frac{1}{2}} + \mathbf{q}^{-\frac{1}{2}} \neq 0$. Therefore $y \underset{L,\varphi}{\leq} w$ and $y \underset{L,\varphi}{\sim} w$. But y and w are not in the same left cell of W (in the sense of 1.11).

Let (W, S) be a Coxeter group. The example leads to the following questions.

(a). Find out all admissible pairs (Σ, φ) (see 1.14) such that every generalized left (resp. two-sided) cell of W with respect to (Σ, φ) is contained in a left (resp. two-sided) cell of W (in the sense of 1.11).

(b). When two admissible pairs give rise to the same generalized left (resp. two-sided) cells of W?

Finally we state

1.23. Conjecture. *Let W be an extended Coxeter group. Let φ and H_φ be as in 1.14-15. Then the generalized left cells and generalized two-sided cells of W defined in 1.15 only depend on the order relations among $\mathbf{q}_s^{\frac{1}{2}}$, $s \in S$, not depend on the parameters $\mathbf{q}_s^{\frac{1}{2}}$, $s \in S$.*

17

2. Affine Weyl groups and Affine Hecke Algebras

In this chapter we consider some elementary properties of extended affine Weyl groups and of their Hecke algebras. We are mainly interested in the structures of the centers of affine Hecke algebras and cells in affine Weyl groups.

2.1. Extended affine Weyl group. Let G be a connected reductive group over $\mathbb{C}$ and T a maximal torus of G. Let $N_G(T)$ be the normalizer of T in G. Then $W_0 = N_G(T)/T$ is a Weyl group, which acts on the character group $X = \text{Hom}(T, \mathbb{C}^*)$ of T. Consider the semi-direct product $W = W_0 \ltimes X$. Let $R \subset X$ be the root system of W_0, which spans the root lattice P in X. The group P is W_0-stable and the subgroup $W' = W_0 \ltimes P$ of W is an affine Weyl group. Moreover W' is a normal subgroup of W. Denote S the set of simple reflections in W'. There exists an abelian subgroup Ω of W such that $\omega S \omega^{-1} = S$ for any $\omega \in \Omega$ and $W = \Omega \ltimes W'$. (Ω is isomorphic to the center of G, also isomorphic to the quotient group X/P.) Thus W is an extended Coxeter group in the sense of 1.1. We shall call W an extended affine Weyl group. (In some bibliography W is also called a modified affine Weyl group, see for example [Ca, 3.5].) The Hecke algebras of W are called affine Hecke algebras. This slightly differs from the usual definition, when G is adjoint, our definition agrees with the usual one.

Let R^+ and R^- be the set of positive roots and the set of negative roots of R respectively. Let Δ be the set of simple roots of R and $R^\vee \subset \text{Hom}(\mathbb{C}^*, T)$ be the dual root system of R. For every $\alpha \in R$ we shall denote $\alpha^\vee \in R^\vee$ the dual of α. Then the length of an element wx ($w \in W_0$, $x \in X$) is given by the following formula (see [IM])

$$(2.1.1) \qquad l(wx) = \sum_{\substack{\alpha \in R^+ \\ w(\alpha) \in R^-}} |\langle x, \alpha^\vee \rangle + 1| + \sum_{\substack{\alpha \in R^+ \\ w(\alpha) \in R^+}} |\langle x, \alpha^\vee \rangle|.$$

Let

$$X^+ = \{x \in X \mid l(wx) = l(w) + l(x) \text{ for any } w \in W_0\} = \{x \in X \mid \langle x, \alpha^\vee \rangle \geq 0\}$$

be the set of dominant weights. By the formula (2.1.1) we get

(2.1.2) For any $x, y \in X^+$, $l(xy) = l(x) + l(y)$.

2.2. The center of generic affine Hecke algebra. Let H be the generic Hecke algebra of W over $A = \mathbb{Z}[\mathbf{q}^{\frac{1}{2}}, \mathbf{q}^{-\frac{1}{2}}]$ with the standard basis T_w, $w \in W$. Following Berstein (see [L5]) we introduced another basis of H and describe the structure of the center of the Hecke algebra.

Given $x \in X$, we can find y, z in X^+ such that $x = yz^{-1}$. Set

$$\theta_x = \mathbf{q}^{-\frac{l(y)}{2}} T_y (\mathbf{q}^{-\frac{l(z)}{2}} T_z)^{-1}.$$

18

According to (2.1.2) we know that θ_x is well defined and is independent of the choice of y, z. Moreover we have

(2.2.1) $\theta_x \theta_{x'} = \theta_{xx'}$,   for any $x, x' \in X$.

Let O_x be the conjugacy class of W containing x and set $z_x = \sum_{x' \in O_x} \theta_{x'}$. Then (Berstein, see [L5])

(a) The elements $T_w \theta_x$, $w \in W_0$, $x \in X$, form an A-basis of H.

(b) The elements $\theta_x T_w$, $w \in W_0$, $x \in X$, form an A-basis of H.

(c) The center of H is a free A-module and the elements z_x, $x \in X^+$, form an A-basis of the center of H.

For $x \in X^+$ and $x' \in X$, we denote $d(x', x)$ the dimension of the x'-weight space $V(x)_{x'}$ of $V(x)$, where $V(x)$ is an irreducible module of G with highest weight x. Set $U_x := \sum_{x' \in X^+} d(x', x) z_{x'}$, $x \in X^+$. Then we have

(d) The elements U_x, $x \in X^+$, form an A-basis of the center of H.

We shall denote S_0 the set of simple reflections in W_0. For each $r \in S_0$, we also write α_r for the corresponding simple root. We have

(e) Let $r \in S_0$ and $x \in X$. Assume that $\langle x, \alpha_r^\vee \rangle = n \geq 1$, then

$$T_r \theta_{r(x)} = \theta_x T_r - (\mathbf{q} - 1)\theta_x (1 + \theta_{\alpha_r}^{-1} + \cdots + \theta_{\alpha_r}^{1-n}).$$

(f) $T_r \theta_x = \theta_x T_r$ for any $r \in S_0$, $x \in X$ with $\langle x, \alpha_r^\vee \rangle = 0$.

A special case of the formula in (e) is

(g) $T_r \theta_{r(x)} = \theta_x T_r - (\mathbf{q} - 1)\theta_x$,   for any $r \in S_0$, $x \in X$ with $\langle x, \alpha_r^\vee \rangle = 1$.

The formula in (g) is equivalent to

(h) $\theta_{r(x)} = \mathbf{q} T_r^{-1} \theta_x T_r^{-1}$,   for any $r \in S_0$, $x \in X$ with $\langle x, \alpha_r^\vee \rangle = 1$.

Given $q \in \mathbb{C}^*$, we can regard $\mathbb{C}$ as an A-algebra through the ring homomorphism $A \to \mathbb{C}$, $\mathbf{q}^{\frac{1}{2}} \to q^{\frac{1}{2}}$ (here $q^{\frac{1}{2}}$ is a square root of q). Then we consider the tensor product $H \otimes_A \mathbb{C}$, which is a $\mathbb{C}$-algebra, denote by $\mathbf{H}_q$. For simplicity we shall denote the images in $\mathbf{H}_q$ of T_w, C_w, U_x, ..., by the same notations respectively.

2.3. Assume that G has a simply connected derived group. The assumption is equivalent to that X^+ contains all fundamental weights concerned with the root system R. For each simple root α in R, we denote x_α the corresponding fundamental weight in X^+. Given $w \in W_0$, set

$$x_w = w\left(\prod_{\substack{\alpha \in \Delta \\ w(\alpha) \in R^-}} x_\alpha \right).$$

19

It is known that (see [St2])

(a) $A[X]$ is a free $A[X]^{W_0}$-module with a basis x_w, $w \in W_0$.

Using 2.2(a-c) we see that

(b) The Hecke algebra H is a free $\mathcal{Z}(H)$-module with a basis $T_w \theta_{x_u}$, $w, u \in W_0$, where $\mathcal{Z}(H)$ is the center of H.

The image in $\mathbf{H}_q$ of $\mathcal{Z}(H)$ is in the center $\mathcal{Z}(\mathbf{H}_q)$ of $\mathbf{H}_q$. By (b) we see that (cf. [M])

(c) Every simple $\mathbf{H}_q$-module has dimension $\leq |W_0|$.

However, for each $q \in \mathbb{C}^*$, there are a lot of simple $\mathbf{H}_q$-modules with dimension $|W_0|$ (see [M, Ka1], see also 3.11). So we have

2.4. Proposition. *Let G be as in 2.3. Then the elements U_x, $x \in X^+$ form a $\mathbb{C}$-basis of the center of $\mathbf{H}_q$. (Recall the convention on notations at the end of 2.2.)*

2.5. Proposition. *Assume that G is a simply connected, simple algebraic group over $\mathbb{C}$. We have*

(i). *If R is not of type D_{2k} ($2k = n$), then there exists a fundamental weight, denoted x_0, such that H is generated by all T_r ($r \in S_0$) and θ_{x_0}.*

(ii). *If R is of type D_{2k} ($2k = n$), then there exist two fundamental weights, denoted x_0, x_0', such that H is generated by all T_r ($r \in S_0$) and $\theta_{x_0}, \theta_{x_0'}$.*

Proof. We number the simple reflections $r_1, r_2, ..., r_n$ in S_0 and the fundamental weights $x_1, x_2, ..., x_n$ in X according to the Coxeter graphs in 1.3. In case (i) we choose x_0 to be the following:

$$x_n, \quad \text{for type } A_n, B_n, D_n, E_n, F_4;$$

$$x_1, \quad \text{for type } C_n, G_2.$$

In case (ii) we choose $x_0 = x_n$ and $x_0' = x_{n-1}$.

We claim that such choices satisfy our requirement. We take type A_n as an example to prove the claim. Let $\underline{H}$ be the subalgebra of H generated by T_r ($r \in S_0$) and $\theta_{x_0} = \theta_{x_n}$. It is enough to prove that the elements $\theta_{x_1}^{\pm 1}, \theta_{x_2}^{\pm 1}, ..., \theta_{x_n}^{\pm 1}$ are in $\underline{H}$. Write T_i for T_{r_i}, and T_i' for $T_{r_i}^{-1}$, respectively. Then we have

(2.5.1) $$r_n(x_n) = x_{n-1} x_n^{-1}, \quad r_1(x_1) = x_2 x_1^{-1},$$

(2.5.2) $$r_i(x_i x_{i+1}^{-1}) = x_{i-1} x_i^{-1}, \quad \text{for } i = 2, 3, ..., n - 1,$$

Using 2.2(h) and (2.5.1) we get

(2.5.3) $$\theta_{x_{n-1}} \theta_{x_n}^{-1} = \mathbf{q} T_n' \theta_{x_n} T_n'.$$

Using 2.2(h) and (2.5.2) repeatedly we obtain

(2.5.4)
$$\theta_{x_{n-2}}\theta_{x_{n-1}}^{-1} = qT'_{n-1}\theta_{x_{n-1}}\theta_{x_n}^{-1}T'_{n-1},$$

$$\cdots\cdots,$$

$$\theta_{x_2}\theta_{x_3}^{-1} = qT'_3\theta_{x_3}\theta_{x_4}^{-1}T'_3,$$

$$\theta_{x_1}\theta_{x_2}^{-1} = qT'_2\theta_{x_2}\theta_{x_3}^{-1}T'_2.$$

Again using 2.2(h) and (2.5.1) we get

(2.5.5)
$$\theta_{x_1}^{-1} = qT'_1\theta_{x_1}\theta_{x_2}^{-1}T'_1.$$

By assumption we have $\theta_{x_n} \in \underline{H}$. Combine this and (2.5.3-5) we see that the elements $\theta_{x_1}^{\pm1}, \theta_{x_2}^{\pm1}, ..., \theta_{x_n}^{\pm1}$ are in $\underline{H}$. The proposition is proved for type A_n. For other types, the arguments are similar.

2.6. Cells in W. In [L11-L14] Lusztig proved a number of results concerned with the cells in an extended affine Weyl group W. Here are some of them except those other references are given.

(a) The number of left cells of W is finite. Each left cell of W contains a unique element of $\mathcal{D} = \{w \in W' \mid 2\deg P_{e,w} = \ell(w) - a(w)\}$, where e is the neutral element of W.

(b) The set $\mathcal{D}$ is a finite set of involutions in W'.

(c) For every element w in W, we have $a(w) \leq a(w_0) = |R^+| = \nu$, where w_0 is the longest element of W_0.

(d) The set $c_0 = \{w \in W \mid a(w) = \nu\}$ is a two-sided cell of W. The two-sided cell c_0 contains $|W_0|$ left (resp. right) cells (see [Bé2, Sh2-Sh3]). The two-sided cell c_0 is the lowest one in the set Cell(W) of two-sided cells of W (concerned with the partial order $\underset{LR}{\leq}$). Assume that X contains all fundamental weights, then

$$c_0 = \{xw_0y \in W \mid x, y \in W \text{ and } l(xw_0y) = l(x) + l(w_0) + l(y)\}.$$

We would like to state a conjecture concerned with the number of left cells in a two-sided cell c. For any subset I of S, let

$$\Gamma_{c,I} = \{\text{left cells } \Gamma \text{ in } c \mid R(\Gamma) = I\},$$

where $R(\Gamma) = R(w)$ for any $w \in \Gamma$, this is well defined, see [KL1]. We conjecture that $\#\Gamma_{c,I} \leq \#\Gamma_{c_0,I}$. Lusztig has a conjecture (see [As]) concerned with the number of left cells in a two-sided cell, which needs the following result (e).

(e) There exists a natural bijection between the set Cell(W) of two-sided cells of W and the set of nilpotent G-orbits in the Lie algebra $\mathbf{g}$ of G.

(f) For $y, w \in W$, we have $a(w) \geq a(y)$ whenever $y \underset{LR}{\leq} w$. Moreover, if $y \underset{LR}{\leq} w$ (resp. $y \underset{L}{\leq} w$, $y \underset{R}{\leq} w$) and $a(y) = a(w)$, then $y \underset{LR}{\sim} w$ (resp. $y \underset{L}{\sim} w$, $y \underset{R}{\sim} w$). In particular, the function a is constant on a two-sided cell of W.

(g) Let x and w be elements in W. Then $x \underset{L}{\leq} w$ (resp. $x \underset{R}{\leq} w$; $x \underset{LR}{\leq} w$) if and only if there exists h (resp. h'; h, h') in H such that $a_x \neq 0$ (resp. $b_x \neq 0$; $c_x \neq 0$), where a_x (resp. b_x; c_x) is defined by

$$hC_w = \sum_{u \in W} a_u C_u, \quad a_u \in A,$$

(resp. $\quad C_w h' = \sum_{u \in W} b_u C_u, \quad b_u \in A; \quad hC_w h' = \sum_{u \in W} c_u C_u, \quad c_u \in A).$

Let c be a two-sided cell of W and u an element in c. We denote $H^{\leq c}$ (resp. $H^{<c}$) the free A-module spanned by C_w, $w \in W$ and $w \underset{LR}{\leq} u$ (resp. $w \underset{LR}{\leq} u$, $w \notin c$). Let $\mathbf{H}_q^{\leq c}$ be the subspace of $\mathbf{H}_q$ spanned by the image of $H^{\leq c}$. By (g) we see that

(h) $H^{\leq c}$ is a two-sided ideal of H and $\mathbf{H}_q^{\leq c}$ is a two-sided ideal of $\mathbf{H}_q$. We also call them cell ideals. We shall write $H_{\geq c}$ for the quotient ring $H/H^{<c}$.

Since $C_{w_0} = \mathbf{q}^{-\frac{l(w_0)}{2}} \sum_{w \in W_0} T_w$ and $T_s C_{w_0} = C_{w_0} T_s = \mathbf{q} C_{w_0}$ for any s in S_0, we have

(i) $H^{\leq c_0}$ is spanned by $\theta_x C_{w_0} \theta_y$, $x, y \in X$. If X contains all fundamental weights of R, then the elements $\theta_{x_w} C_{w_0} \theta_{x_u} U_x$, $w, u \in W_0$, $x \in X^+$, form an A-basis of $H^{\leq c_0}$.

Proof. Using 2.2(a-b) we get the first assertion. The second assertion follows from 2.3(b) and [X2, 2.9].

2.7. The based ring of two-sided cell. For $w, u, v \in W$, we define the integer $\gamma_{w,u,v}$ by the condition

$$\mathbf{q}^{\frac{a(v)}{2}} h_{w,u,v} - \gamma_{w,u,v} \in \mathbf{q}^{\frac{1}{2}} \mathbb{Z}[\mathbf{q}^{\frac{1}{2}}],$$

see 1.12 for the definition of $h_{w,u,v}$. Since (W', S) is crystallographic, by 1.12 we know that $\gamma_{w,u,v}$ is non-negative (see [L11]). We have

(a) If $\gamma_{w,u,z}$ is not zero, then $w \underset{L}{\sim} u^{-1}$, $u \underset{L}{\sim} z$, and $w \underset{R}{\sim} z$. In particular, we have $w \underset{LR}{\sim} u \underset{LR}{\sim} z$ if $\gamma_{w,u,z}$ is not zero.

(b) Let $d \in \mathcal{D}$, then $\gamma_{w,d,u} \neq 0 \Leftrightarrow w = u$ and $w \underset{L}{\sim} d$, $\gamma_{w,d,w} = 1$. Moreover, $\gamma_{w,d,w} = \gamma_{d,w^{-1},w^{-1}} = \gamma_{w,w^{-1},d} = 1$.

Let $J_\mathbb{Z}$ be the free $\mathbb{Z}$-module with a basis t_w, $w \in W$. In [L12] Lusztig proved that $t_w t_u = \sum_{v \in W} \gamma_{w,u,v} t_v$ defines an associative ring structure on $J_\mathbb{Z}$. The ring $J_\mathbb{Z}$

22

is called the based ring of W. The unit in $J_{\mathbb{Z}}$ is $\sum_{d \in \mathcal{D}} t_d$. According to (a), for each two-sided cell c, the subspace $J_{\mathbb{Z},c}$ of $J_{\mathbb{Z}}$ spanned by t_w, $w \in c$, is a two-sided ideal of $J_{\mathbb{Z}}$. The ideal $J_{\mathbb{Z},c}$ is in fact an associative ring with unit $\sum_{d \in \mathcal{D} \cap c} t_d$, which is called the based ring of the two-sided cell c. We have $J_{\mathbb{Z}} = \oplus_c J_{\mathbb{Z},c}$, i.e. $J_{\mathbb{Z}}$ is the direct sum of the rings $J_{\mathbb{Z},c}$, where c runs over the set $\mathrm{Cell}(W)$ of two-sided cells of W. The rings $J_{\mathbb{Z}}$ and $J_{\mathbb{Z},c}$ turn out to be very interesting. The following result establishs the connection between the Hecke algebra H and the based ring $J_{\mathbb{Z}}$.

(c) The A-linear map $\phi \colon H \to J_{\mathbb{Z}} \otimes_{\mathbb{Z}} A$ defined by

$$\phi(C_w) = \sum_{\substack{u \in W \\ d \in \mathcal{D} \\ a(d) = a(u)}} h_{w,d,u} t_u, \qquad w \in W$$

is a homomorphism of A-algebras with unit. Moreover ϕ is injective and ϕ maps the center of H into the center of $J_{\mathbb{Z}} \otimes_{\mathbb{Z}} A$ and $J_{\mathbb{Z}} \otimes_{\mathbb{Z}} A$ is finitely generated over $\phi(\mathcal{Z}(H))$. (See [L13]).

The $\mathbb{C}$-algebra $\mathbf{J} = J_{\mathbb{Z}} \otimes \mathbb{C}$ is called the asymptotic Hecke algebra of W. The homomorphism ϕ in (c) induces a homomorphism of $\mathbb{C}$-algebras $\phi_q : \mathbf{H}_q \to \mathbf{J}$. We have (see [L13])

(d) For any $q \in \mathbb{C}^*$, the homomorphism ϕ_q is injective and ϕ_q maps the center of $\mathbf{H}_q$ into the center of $\mathbf{J}$, and $\mathbf{J}$ is finitely generated over $\phi_q(\mathcal{Z}(\mathbf{H}_q))$.

We conjecture that the center of $\mathbf{J}$ is spanned by various $\phi_q(\mathcal{Z}(\mathbf{H}_q))$, $q \in \mathbb{C}^*$.

Affine Hecke algebras with two parameters

2.8. We keep the notations in 2.1 and consider the affine Hecke algebras with unequal parameters. In 2.8-2.12 we assume that G is simple and G is one of the following types: B_n, C_n, F_4, G_2. Thus there exist simple reflections in S which are not conjugate in W.

Let R' be the set of long roots in R provided that R is of type B_n, and be the set of short roots in R provided that R is one of other types. Let $R'' = R - R'$ and set $R'^+ = R' \cap R^+$(resp. $R''^+ = R'' \cap R^+$) be the set of positive roots in R' (resp. R''). Define two functions l', $l'' : W \to \mathbb{N}$ as follows. Given $w \in W_0$ and $x \in X$, we set

(2.8.1)
$$l'(wx) = \sum_{\substack{\alpha \in R'^+ \\ w(\alpha) \in R^-}} |\langle x, \alpha^\vee \rangle + 1| + \sum_{\substack{\alpha \in R'^+ \\ w(\alpha) \in R^+}} |\langle x, \alpha^\vee \rangle|.$$

(2.8.2)
$$l''(wx) = \sum_{\substack{\alpha \in R''^+ \\ w(\alpha) \in R^-}} |\langle x, \alpha^\vee \rangle + 1| + \sum_{\substack{\alpha \in R''^+ \\ w(\alpha) \in R^+}} |\langle x, \alpha^\vee \rangle|.$$

In some sense the functions l', l'' are length functions of W corresponding to roots of different lengths. To see it we introduce some notations. Define

$$S' = \{s \in S \mid s \text{ is the simple reflection with respect to some roots in } R'\},$$

$$S'' = \{s \in S \mid s \text{ is the simple reflection with respect to some roots in } R''\}.$$

For any reduced expression $t_1 t_2 \cdots t_k$ of an element u in W', set

$$S'_u = \{t_i \mid t_i \in S' \text{ and } 1 \leq i \leq k\},$$

$$S''_u = \{t_i \mid t_i \in S'' \text{ and } 1 \leq i \leq k\},$$

then $l'(u) = \#S'_u$ and $l''(u) = \#S''_u$. Obviously we have

(2.8.3). Let $u \in W$, then $l(u) = l'(u) + l''(u)$. (See (2.1.1) for the formula of $l(u)$).

Let $\mathbf{u}, \mathbf{v}$ be two indeterminates and let $B = \mathbb{Z}[\mathbf{u}^{\pm 1}, \mathbf{v}^{\pm 1}]$ be the ring of all Laurant polynomials in $\mathbf{u}, \mathbf{v}$ with integer coefficients. We define the Hecke algebra $\tilde{H}$ (over B) of W with parameters $\mathbf{u}^2, \mathbf{v}^2$ as follows: $\tilde{H}$ is a free B-module with a basis T_w, $w \in W$ and the multiplication laws are

(2.8.4)
$$T_w T_u = T_{wu}, \quad \text{if } l(wu) = l(w) + l(u),$$

$$(T_s - \mathbf{u}^2)(T_s + 1) = 0, \quad \text{if } s \in S',$$

$$(T_s - \mathbf{v}^2)(T_s + 1) = 0, \quad \text{if } s \in S''.$$

When G is not adjoint, or is not of type B_n, the algebra $\tilde{H}$ is essentially the generic Hecke algebra $\mathcal{H}$ of W in the sense of 1.1. When G is adjoint and is of type B_n as well, then $W = W'$ is an affine Weyl group of type $\tilde{C}_n$. We number the simple reflections in S according to the Coxeter graph in 1.3, then s_n, s_0 are not conjugate in W. So the generic Hecke algebra $\mathcal{H}$ of W has three parameters. In this book we donot consider such a Hecke algebra although the properties for $\tilde{H}$ have their counterparts for the algebra $\mathcal{H}$ (see [L18]). It seems necessary to consider the Hecke algebra $\mathcal{H}$ separatedly.

2.9. The center of $\tilde{H}$. Following [L18] we construct the center of $\tilde{H}$, which essentially is similar to the case in [L18].

As in 2.1, let

$$X^+ = \{x \in X \mid l(wx) = l(w) + l(x) \text{ for any } w \in W_0\} = \{x \in X \mid \langle x, \alpha^\vee \rangle \geq 0\}$$

be the set of dominant weights. By the formulas (2.8.1-2) we have

(2.9.1) Let $x, y \in X^+$, then $l'(xy) = l'(x) + l'(y)$ and $l''(xy) = l''(x) + l''(y)$.

For every $x \in X$, we can find y, z in X^+ such that $x = yz^{-1}$. Let

$$\theta_x = \mathbf{u}^{-l'(y)} \mathbf{v}^{-l''(y)} T_y (\mathbf{u}^{-l'(z)} \mathbf{v}^{-l''(z)} T_z)^{-1}.$$

24

According to (2.9.1) we know that θ_x is well defined and is independent of the choice of y, z. Moreover we have

(2.9.2) $$\theta_x \theta_{x'} = \theta_{xx'}, \quad \text{for any } x, x' \in X.$$

Let O_x be the conjugacy class of W containing x and let $z_x = \sum_{x' \in O_x} \theta_{x'}$. Then using the method in [L18] one can prove the following result without difficulty.

(a) The elements $T_w \theta_x$, $w \in W_0$, $x \in X$, form a B-basis of $\tilde{H}$.

(b) The elements $\theta_x T_w$, $w \in W_0$, $x \in X$, form a B-basis of $\tilde{H}$.

(c) The center of $\tilde{H}$ is a free B-module and the elements z_x, $x \in X^+$, form a B-basis of the center of $\tilde{H}$.

For $x \in X^+$ and $x' \in X$, we set $U_x = \sum_{x' \in X^+} d(x', x) z_{x'}$, $x \in X^+$ (see 2.2 for the definition of $d(x', x)$). Then we have

(d) The elements U_x, $x \in X^+$, form a B-basis of the center of $\tilde{H}$.

Define $S_0 = S \cap W_0$, $S_0' = S' \cap W_0$ and $S_0'' = S'' \cap W_0$.

(e) Let $r \in S_0'$ and $x \in X$. Assume that $\langle x, \alpha_r^\vee \rangle = n \geq 1$. Then

$$T_r \theta_{r(x)} = \theta_x T_r - (u^2 - 1)\theta_x(1 + \theta_{\alpha_r}^{-1} + \cdots + \theta_{\alpha_r}^{1-n}).$$

Let $r \in S_0''$ and $x \in X$. Assume that $\langle x, \alpha_r^\vee \rangle = n \geq 1$. Then

$$T_r \theta_{r(x)} = \theta_x T_r - (v^2 - 1)\theta_x(1 + \theta_{\alpha_r}^{-1} + \cdots + \theta_{\alpha_r}^{1-n}).$$

(f) $T_r \theta_x = \theta_x T_r$ \qquad for any $r \in S_0$, $x \in X$ with $\langle x, \alpha_r^\vee \rangle = 0$.

A special case of the formulas in (e) are

(g) $T_r \theta_{r(x)} = \theta_x T_r - (u^2 - 1)\theta_x$, for any $r \in S_0'$, $x \in X$ with $\langle x, \alpha_r^\vee \rangle = 1$.

$\quad T_r \theta_{r(x)} = \theta_x T_r - (v^2 - 1)\theta_x$, for any $r \in S_0''$, $x \in X$ with $\langle x, \alpha_r^\vee \rangle = 1$.

The formulas in (g) are equivalent to

(h) $\theta_{r(x)} = u^2 T_r^{-1} \theta_x T_r^{-1}$, for any $r \in S_0'$, $x \in X$ with $\langle x, \alpha_r^\vee \rangle = 1$.

$\quad \theta_{r(x)} = v^2 T_r^{-1} \theta_x T_r^{-1}$, for any $r \in S_0''$, $x \in X$ with $\langle x, \alpha_r^\vee \rangle = 1$.

We shall denote $\tilde{H}^\circ$ the $B' = \mathbb{Z}[u^{\pm 2}, v^{\pm 2}]$-subalgebra of $\tilde{H}$ generated by T_s, θ_x, $s \in S_0$, $x \in X$.

Given $a, b \in \mathbb{C}^*$, we can regard $\mathbb{C}$ as a B-algebra through the ring homomorphism $B \to \mathbb{C}$, $u \to a^{\frac{1}{2}}$, $v \to b^{\frac{1}{2}}$ (here $a^{\frac{1}{2}}$ and $b^{\frac{1}{2}}$ are square roots a and b respectively). Then we consider the tensor product $\tilde{H} \otimes_B \mathbb{C}$, which is a $\mathbb{C}$-algebra, denote by $\tilde{\mathbf{H}}_{a,b}$. It is easy to see that $\tilde{\mathbf{H}}_{a,b}$ only depends on a, b, does not depend on the choices of the square roots of a, b, actually we have $\tilde{\mathbf{H}}_{a,b} = \tilde{H}^\circ \otimes_{B'} \mathbb{C}$. We shall denote the images in $\tilde{\mathbf{H}}_{a,b}$ of T_w, θ_x, $U_x, \dots$ by the same notations respectively.

2.10. Assume that G is simply connected. For each $w \in W_0$, we define $x_w \in X$ as in 2.3. Using 2.9 (a-c) and 2.3 (a) we see that

(a) $\tilde{H}$ is a free $\mathcal{Z}(\tilde{H})$-module with a basis $T_w \theta_{x_u}$, $w, u \in W_0$, where $\mathcal{Z}(\tilde{H})$ is the center of $\tilde{H}$.

By (a) we see that (see [M])

(b) Any simple $\tilde{\mathbf{H}}_{a,b}$-module has dimension $\leq |W_0|$.

But for any $a, b \in \mathbb{C}^*$, there are a lot of simple $\tilde{\mathbf{H}}_{a,b}$-modules with dimension $|W_0|$ (see [M, Ka1]). So we have

2.11. Proposition. *Assume that X contains all fundamental weights of R, then the elements U_x, $x \in X^+$ form a $\mathbb{C}$-basis of the center of $\tilde{\mathbf{H}}_{a,b}$. (Recall the convention on notations at the end of 2.9.)*

Similar to 2.5 we have

2.12. Proposition. *Assume that G is a simply connected, simple algebraic group over $\mathbb{C}$, and is one of the types B_n, C_n, F_4, G_2. Then there exists a fundamental weight, denoted x_0, such that $\tilde{H}$ is generated by all T_r ($r \in S_0$) and θ_{x_0}.*

2.13. Assume that G is a simply connected, simple algebraic group over $\mathbb{C}$. Let the notation W, S, Ω, W_0, ... be as in 2.1. Let Γ be an abelian group and let $\varphi : W \to \Gamma$ be a map such that $\varphi(\omega s_1 s_2 \cdots s_k \omega') = \varphi(s_1)\varphi(s_2) \cdots \varphi(s_k)$ for any reduced expression $s_1 s_2 \cdots s_k$ in W' and ω, ω' in Ω. For each w in W we shall write $\mathbf{q}_w^{\frac{1}{2}}$ for $\varphi(w)$. Recall that we denote H_φ the Hecke algebra (over $\mathbb{Z}[\Gamma]$) of W with respect to φ. When W is of types $\tilde{A}_n, \tilde{D}_n$ and $\tilde{E}_n$, there is a natural ring homomorphism $H \to H_\varphi$ such that $\mathbf{q}_s^{\frac{1}{2}}$ maps to $\mathbf{q}_s^{\frac{1}{2}}$ ($s \in S$) and T_w maps to T_w ($w \in W$). When W is of types $\tilde{B}_n, \tilde{C}_n, \tilde{F}_4$ and $\tilde{G}_2$, there is a natural ring homomorphism $\tilde{H} \to H_\varphi$ such that $\mathbf{u}$ maps to $\mathbf{q}_s^{\frac{1}{2}}$ ($s \in S'$), $\mathbf{v}$ maps to $\mathbf{q}_s^{\frac{1}{2}}$ ($s \in S''$) and T_w maps to T_w ($w \in W$). (See 1.14, 2.2 and 2.8 for the definitions of H_φ, H, $\tilde{H}$, respectively.)

Convention. The images in H_φ of $\theta_x, U_x,...$, will be denoted by the same notations respectively.

Assume that we have a total order Σ on Γ which is compatible with the multiplication of Γ and further require that all $\mathbf{q}_s^{\frac{1}{2}}$ ($s \in S$) are positive. Then we may define the generalized left (resp. two-sided) cells of W with respect to the pair (Σ, φ). In next chapter we shall prove that the set c_0 in 2.6 (d) is always a generalized two-sided cell of W with respect to the pair (Σ, φ). Of course the result is new only for types $\tilde{B}_n, \tilde{C}_n, \tilde{F}_4$ and $\tilde{G}_2$.

3. A Generalized Two-sided Cell of an Affine Weyl Group

In this chapter we shall prove that the set c_0 in 2.6 (d) is always a generalized two-sided cell (in the sense of 1.15) of an affine Weyl group. For this we need to generalize some results of [L2, L5]. More or less, such generalizations are easy and straight although they seem interesting. Most proofs are similar to those in [L2, L5]. The results in [Ka2, Gu] also can be generalized without much difficulty. We do it in next chapter.

The contents of the chapter are as follows. First we realize an irreducible affine Weyl group as a group of affine motions of an Euclidean space, then generalize some results of [L2, L5]. Finally we prove the main result of the chapter (Theorem 3.22).

3.1. Let V be an Euclidean space and let $R \subset V$ be an irreducible root system of rank $\dim V$. Let $P = \mathbb{Z}R$ be the root lattice of V generated by elements in R. Denote $R^\vee$ the dual root system of R.

We may require the scalar product $(\ ,\ )$ of V satisfying $2(\alpha, \beta) = \langle \alpha, \beta^\vee \rangle (\beta, \beta)$ for any roots α, β in R. Then the scalar product is invariant under the Weyl group W_0 of R, that is, $(wv_1, wv_2) = (v_1, v_2)$ for any w in W_0 and v_1, v_2 in V. We shall identify $\alpha^\vee$ with $2\alpha/(\alpha, \alpha)$ for each root α in R. Then $R^\vee = \{2\alpha/(\alpha, \alpha) \mid \alpha \in R\}$. For $v \in V$ and $\alpha \in R$ we also write $\langle v, \alpha^\vee \rangle$ instead of $(v, \alpha^\vee)$.

The set $X := \{x \in V \mid \langle x, \alpha^\vee \rangle$ is an integer for each α in $R\}$ is W_0-invariant and P is a W_0-invariant subset of X. The semi-direct product $W_a = W_0 \ltimes P$ is an affine Weyl group and the semi-direct product $W = W_0 \ltimes X$ is the extended affine Weyl group associated to a simply connected, simple algebraic group G (over $\mathbb{C}$) of root system R. Moreover W_a is a normal subgroup of W and there exists a finite abelian subgroup Ω of W such that W is isomorphic to the semi-direct product $\Omega \ltimes W_a$. To avoid confusion we denote p_x the element $x \in X$ when it is regarded as an element in W. When X is not regarded as a subgroup of W, the operation in X will be written additively.

We may realize W_a as a group of affine motions of V. Given $\alpha \in R^+$ (the set of positive roots in R) and $n \in \mathbb{Z}$, denote $F_{\alpha, n}$ the hyperplane $\{v \in V \mid \langle v, \alpha^\vee \rangle = n\}$ of V. The corresponding reflection is

$$\sigma_{\alpha, n} = \sigma_{F_{\alpha, n}} : v \to v - (\langle v, \alpha^\vee \rangle - n)\alpha, \qquad v \in V.$$

Then W_a is isomorphic to the group $\mathcal{W}_a$ of affine motions of V generated by the reflections $\sigma_{\alpha, n}$, $\alpha \in R^+$, $n \in \mathbb{Z}$.

The above isomorphism gives rise to another useful realization of W_a which we explain now.

A connected component of $V - \bigcup\limits_{\substack{\alpha \in R^+ \\ n \in \mathbb{Z}}} F_{\alpha, n}$ is called an alcove, which is an open polyhedron of dimension n with $n + 1$ faces (=codimension 1 facets). Let S be the set of $\mathcal{W}_a$-orbits of the set of faces of alcoves. Then S contains $n + 1$ elements which can be represented as the $n+1$ faces of any given alcove. We shall denote Υ the set of alcoves and denote $\mathcal{F}$ the set of the hyperplanes $F_{\alpha,n}$, $\alpha \in R^+$, $n \in \mathbb{Z}$.

For any $s \in S$ and $A \in \Upsilon$, denote sA the unique alcove which shares exactly one face with A and the shared face is of type s. The map $A \to sA$ is an involution. All such involutions generate a group $\mathcal{W}'_a$ of permutations of Υ and $(\mathcal{W}'_a, S)$ is an affine Weyl group. We shall regard $\mathcal{W}'_a$ acting on Υ through left hand side and $\mathcal{W}_a$ acting on Υ through right hand side. The two actions are commutative and both actions are simply transtive.

We may establish the isomorphism between $\mathcal{W}'_a$ and $\mathcal{W}_a$ as follows. Let A be an alcove and w' be an element in $\mathcal{W}'_a$. Then there exists a unique element w in $\mathcal{W}_a$ such that $w'(Ay) = (A)wy$ for any y in $\mathcal{W}_a$. The map $w' \to w^{-1}$ defines an isomorphism between $\mathcal{W}'_a$ and $\mathcal{W}_a$. So $\mathcal{W}'_a$ is isomorphic to $\mathcal{W}_a$. We shall identify $\mathcal{W}_a$ with $\mathcal{W}'_a$ and with $\mathcal{W}_a$, and regard S as the set of simple reflections of $\mathcal{W}_a$. We denote S_0 the set of simple reflections in W_0. We have $(v)p_x = v + x$ for any v in V and x in P.

3.2. A hyperplane $F = F_{\alpha,n}$ divides $V - F$ into two parts

$$V_F^+ := \{v \in V \mid \langle v, \alpha^\vee \rangle > n\}, \quad V_F^- := \{v \in V \mid \langle v, \alpha^\vee \rangle < n\} = \sigma_{\alpha,n} V_F^+.$$

Let A be an alcove. For any simple reflection $s \in S$, we denote $F_{s,A}$ the unique hyperplane separating A from sA (that is, neither $V_{F_{s,A}}^+$ nor $V_{F_{s,A}}^-$ contains $A \cup sA$). Define

$$\mathcal{L}(A) := \{s \in S \mid A \text{ is contained in } V_{F_{s,A}}^+\}.$$

A point v of V is said special if there are $|R^+|$ hyperplanes in $\mathcal{F}$ containing v (recall that R^+ is the set of positive roots of R). For example, v is a special point if v is in the root lattice P. For any special point v, we denote C_v^+ the quarter $\bigcap_{\substack{F \in \mathcal{F} \\ F \ni v}} V_F^+$ and

let A_v^+ be the unique alcove in C_v^+ whose closure contains v.

Let A_0 be the fundamental alcove

$$\{v \in V \mid 0 < \langle v, \alpha^\vee \rangle < 1 \text{ for every positive root } \alpha\}$$

and let A_0^- be the negative fundamental alcove

$$\{v \in V \mid -1 < \langle v, \alpha^\vee \rangle < 0 \text{ for every positive root } \alpha\}.$$

For each element x in the root lattice $P = \mathbb{Z}R$ we have $A_x^+ = A_0 p_x$. For later uses we set $A_x^- := A_0^- p_x$. (Recall that P is regarded as a subgroup of $\mathcal{W}_a$ and $\mathcal{W}_a$ acts on Υ through right hand side.)

Let A, B be two alcoves. We associate to A, B an integer $d(A, B)$ as follows. There are only finitely many hyperplanes F in $\mathcal{F}$ separating A from B (i.e., such that A, B are on different sides of F). We count these hyperplanes with alternating signs. A hyperplane F is counted with $+1$ if $A \subset V_F^-$ and $B \subset V_F^+$; is counted with -1 if $A \subset V_F^+$ and $B \subset V_F^-$. The sum of these ± 1 over all hyperplanes F in $\mathcal{F}$ separating A from B is denoted $d(A, B)$.

We now introduce a partial order $\leq$ on Υ. We say that $A \leq B$ if there exists a sequence of alcoves $A = A_1, A_2, ..., A_k = B$ such that $d(A_i, A_{i+1}) = 1$ and $A_{i+1} = A_i \sigma_{F_i}$ for $i = 1, 2, ..., k - 1$ and some $F_i \in \mathcal{F}$.

3.3. Let φ be a map from the extended affine Weyl group W to an abelian group Γ such that $\varphi(\omega s_1 s_2 \cdots s_k \omega') = \varphi(s_1)\varphi(s_2)\cdots\varphi(s_k)$ for any reduced expression $s_1 s_2 \cdots s_k$ in W_a and ω, ω' in Ω. We shall write $\mathbf{q}_w^{\frac{1}{2}}$ rather than $\varphi(w)$ for each w in W. Note that $\varphi(w) = \varphi(w')$ whenever w, w' are conjugate in W. Let β be a simple root and let s be the corresponding simple reflection. Assume that α is root and $\alpha = w(\beta)$ for some w in W_0. For convenience we set

$$\mathbf{q}_\alpha^{\frac{1}{2}} = \mathbf{q}_s^{\frac{1}{2}}, \qquad \mathbf{q}_F^{\frac{1}{2}} = \mathbf{q}_\alpha^{\frac{1}{2}} \text{ if } F = F_{\alpha,n} \text{ for some integer } n.$$

Recall that the Hecke algebra H_φ of W with respect to φ is an algebra over the group ring $\mathbb{Z}[\Gamma]$ (see 1.14). As a $\mathbb{Z}[\Gamma]$-module, it is free with a basis T_w, $w \in W$. The multiplication is defined by

$$(T_s - \mathbf{q}_s)(T_s + 1) = 0, \quad \text{if } s \in S; \qquad T_w T_u = T_{wu}, \quad \text{if } l(wu) = l(w) + l(u).$$

Denote H'_φ be the subalgebra of H_φ generated by T_s, $s \in S$.

Let $\mathcal{M}$ be the free $\mathbb{Z}[\Gamma]$-module with a basis $A, A \in \Upsilon$. There is a unique H'_φ-module structure on $\mathcal{M}$ such that for any A in Υ and s in S we have

$$(3.3.1) \qquad T_s A = \begin{cases} sA, & \text{if } s \notin \mathcal{L}(A), \\ \mathbf{q}_s sA + (\mathbf{q}_s - 1)A, & \text{if } s \in \mathcal{L}(A). \end{cases}$$

(The verification of the definition relations in H'_φ is immediate, compare with [M, 4.1.1; L2, 1.6].)

3.4. Lemma. *Let v be a special point and let A be an alcove such that its closure $\bar{A}$ contains v. Let $y \in W_a$ be such that $y(A_v^+) \subset C_v^+$ and write*

$$T_y(A) = \sum_{B \in \Upsilon} \pi_{B, y(A_v^+)}^A B \in \mathcal{M}.$$

Then

(i). $\pi_{B, y(A_v^+)}^A$ *is a polynomial in $\mathbf{q}_s, s \in S$ if $B \leq y(A_v^+)$ and is zero otherwise.*

(ii). $\pi_{y(A_v^+), y(A_v^+)}^A = \begin{cases} 0, & \text{if } A \text{ is not equal to } A_v^+, \\ 1, & \text{if } A = A_v^+. \end{cases}$

Proof. Let $s_1 s_2 \cdots s_k$ be a reduced expression of y. By (3.3.1) we know that

$$T_y(A) = \sum_I f_I s_1 \cdots \hat{s}_{i_1} \cdots \hat{s}_{i_2} \cdots \hat{s}_{i_p} \cdots s_k(A),$$

where the sum runs through all subsets $I = \{i_1 < i_2 \cdots < i_p\}$ of 1,2,...,k and f_I is a polynomial in $\mathbf{q}_s, s \in S$. As the proof of [L2, Lemma 4.2, p.134] we know that the lemma is true.

29

3.5. Let α be a positive root. The connected components of $V - \bigcup_{n\in\mathbb{Z}} F_{\alpha,n}$ are called strips. We shall write $\partial^+ U$ and $\partial^- U$ for $F_{\alpha,n+1}$ and $F_{\alpha,n}$ respectively when U is a strip between $F_{\alpha,n+1}$ and $F_{\alpha,n}$. For later uses we define (compare with [L2, 2.3])

$$d(F_{\alpha,m}, F_{\alpha,n}) := n - m, \qquad d(F_{\alpha,m}, v) := n - m,$$

where v is any special point in $F_{\alpha,n}$.

Let $\mathcal{M}_\alpha$ be the $\mathbb{Z}[\Gamma]$-submodule of $\mathcal{M}$ spanned by the elements

$$f_A := A + A\sigma_F,$$

where $F \in \mathcal{F}_\alpha := \{F_{\alpha,n} \mid n \in \mathbb{Z}\}$ is determined by the condition $F = \partial^+$ (the strip containing A) and σ_F is the reflection of V corresponding to F (see 3.1). Obviously the elements $f_A, A \in \Upsilon$ constitute a basis of $\mathcal{M}_\alpha$.

3.6. Lemma. *Let α be a simple root and $F \in \mathcal{F}_\alpha$. Then*

(i). $\mathcal{M}_\alpha$ *is an H'_φ-submodule of $\mathcal{M}$.*

(ii). *The $\mathbb{Z}[\Gamma]$-linear map*

$$\kappa_F : \mathcal{M}_\alpha \to \mathcal{M}_\alpha, \quad A + A\sigma_{F_1} \to \mathbf{q}_\alpha^{d(F,F_1)}(A\sigma_F + A\sigma_{F_1}\sigma_F),$$

is a homomorphism of H'_φ-module, where $F_1 = \partial^+$ (the strip containing A).

(Compare with [L2, Lemma 2.4].)

Proof. Set $A' = A\sigma_{F_1}$. Then $A\sigma_F + A'\sigma_F = f_{A'\sigma_F} \in \mathcal{M}_\alpha$. Let s be an element in S and let L be the hyperplane separating A from sA.

Case 1. L is not in $\mathcal{F}_\alpha$. It follows that

(a). The hyperplane separating A' from sA' is not in $\mathcal{F}_\alpha$;

(b). sA and A are in the same strip;

(c). sA' and A' are in the same strip;

(d). The four conditions: $s \in \mathcal{L}(A)$, $s \in \mathcal{L}(A')$, $s \in \mathcal{L}(A\sigma_F)$, $s \in \mathcal{L}(A'\sigma_F)$ are equivalent since $V_L^+ \sigma_F = V_{L\sigma_F}^+$ and $V_L^- \sigma_F = V_{L\sigma_F}^-$. (Note that α is a simple root.)

These imply that $sA + sA' \in \mathcal{M}_\alpha$ and

$$T_s(A + A') = \begin{cases} sA + sA', & \text{if } s \notin \mathcal{L}(A), \\ \mathbf{q}_s(sA + sA') + (\mathbf{q}_s - 1)(A + A'), & \text{if } s \in \mathcal{L}(A). \end{cases}$$

If $s \notin \mathcal{L}(A)$, we have

$$(T_s(A + A'))\kappa_F = T_s((A + A')\kappa_F) = \mathbf{q}_\alpha^{d(F,F_1)}(sA\sigma_F + sA'\sigma_F).$$

If $s \in \mathcal{L}(A)$, we have

$$(T_s(A + A'))\kappa_F = T_s((A + A')\kappa_F)$$
$$= \mathbf{q}_\alpha^{d(F,F_1)}(\mathbf{q}_s(sA\sigma_F + sA'\sigma_F) + (\mathbf{q}_s - 1)(A\sigma_F + A'\sigma_F)).$$

30

Case 2. $L = F_1$. It follows that $A' = sA$. We have

$$T_s(A + A') = \mathbf{q}_s(A + A'),$$

$$(T_s(A + A'))\kappa_F = T_s((A + A')\kappa_F) = \mathbf{q}_\alpha^{d(F,F_1)}\mathbf{q}_s(A\sigma_F + A'\sigma_F).$$

Case 3. $L \in \mathcal{F}_\alpha$ but L is not equal to F_1. It follows that $L = \partial^-$ (the strip containing A). Then we have $\mathbf{q}_\alpha = \mathbf{q}_s$. Let $A'' = A\sigma_L = sA$. Direct computation shows that

$$T_s f_A = \mathbf{q}_s f_{A''} - f_A + f_{A'},$$

$$(T_s f_A)\kappa_F = T_s(f_A\kappa_F) = \mathbf{q}_\alpha^{d(F,F_1)}(f_{A''\sigma_F} - f_{A\sigma_F} + \mathbf{q}_s f_{A'\sigma_F}).$$

The lemma is proved.

3.7. Let v be a special point. Set

$$e_v = \sum_{\substack{A \\ \bar{A} \ni v}} A \in \mathcal{M},$$

and let $\mathcal{M}_v$ be the H'_φ-submodule of $\mathcal{M}$ generated by e_v. Let α be a simple root and n be an integer such that $F = F_{\alpha,n}$ contains v. Let $\mathcal{W}_F$ be the dihedral subgroup of W generated by the reflections corresponding to the hyperplanes in $\mathcal{F}_\alpha$. Let A be an alcove between $F_{\alpha,i+1}$ and $F_{\alpha,i}$. For each integer j in $\mathbb{Z}$ we denote $A^{(j)}$ the unique alcove in $A\mathcal{W}_F$ such that $A^{(j)}$ is between $F_{\alpha,i+j+1}$ and $F_{\alpha,i+j}$ if $i \geq n$, is between $F_{\alpha,i-j+1}$ and $F_{\alpha,i-j}$ if $i < n$. Obviously $(A\sigma_F)^{(j)} = (A^{(j)})\sigma_F$. Note that $d(A^{(a)}, A^{(b)}) = b - a$ if $A \subset V_F^+$ (see [L2, Lemma 2.5, p.127]).

Let $f = \sum \xi_A A \in \mathcal{M}$. We define a new element $f^* = \sum \xi_A^* A \in \mathcal{M}$ by the formula

$$\xi_A^* = \xi_{A^{(0)}} - \xi_{A^{(1)}} + \xi_{A^{(2)}} - \cdots.$$

Clearly f^* is well defined.

3.8. Lemma. *Let v be a special point and α be a simple root. Assume that $F \in \mathcal{F}_\alpha$. Then*

(i). $e_v \in \mathcal{M}_\alpha$ and $e_v\kappa_F = \mathbf{q}_\alpha^{d(F,v)} e_{v\sigma_F}$.

(ii). *Assume that F contains v and f is an element in $\mathcal{M}_v$. Let A be an alcove in V_F^+. Then*

$$\xi_{A\sigma_F}^* = \mathbf{q}_\alpha^{\frac{d(A\sigma_F,A)-1}{2}} \xi_A^*.$$

(Compare with [L2, Lemma 2.6 and Prop. 9.2].)

Proof. (i). Let F' be the unique hyperplane in $\mathcal{F}_\alpha$ containing v. We have

$$e_v = \sum_{\substack{A \subset V_{F'}^- \\ \bar{A} \ni v}} f_A$$

31

and

$$e_v \kappa_F = \sum_{\substack{A \subset V_{F'}^- \\ \bar{A} \ni v}} \mathbf{q}_\alpha^{d(F,F')} f_{A\sigma_{F'}\sigma_F} = \mathbf{q}_\alpha^{d(F,F')} \sum_{\substack{A' \subset V_{F'\sigma_F}^- \\ \bar{A}' \ni v\sigma_F}} f_{A'} = \mathbf{q}_\alpha^{d(F,v)} e_{v\sigma_F}.$$

(ii). By (i), f is in $\mathcal{M}_\alpha$. Let $f = \sum_{B \in \Upsilon} \xi_B B = \sum_{B \in \Upsilon} \xi'_B (B + B\sigma_{F_B})$, where $F_B = \partial^+$(the strip containing B). Then

$$\xi_A^* = (\xi'_{A(-1)} + \xi'_{A(0)}) - (\xi'_{A(0)} + \xi'_{A(1)}) + (\xi'_{A(1)} + \xi'_{A(2)}) - \cdots = \xi'_{A(-1)}.$$

Since F contains v, by (i) we get $e_v \kappa_F = e_v$. According to Lemma 3.6 (ii) we have $f \kappa_F = f$. Therefore

$$\xi_{A\sigma_F}^* = (\xi'_{A(-1)} \mathbf{q}_\alpha^d + \xi'_{A(0)} \mathbf{q}_\alpha^{d+1}) - (\xi'_{A(0)} \mathbf{q}_\alpha^{d+1} + \xi'_{A(1)} \mathbf{q}_\alpha^{d+2})$$
$$+ (\xi'_{A(1)} \mathbf{q}_\alpha^{d+2} + \xi'_{A(2)} \mathbf{q}_\alpha^{d+3}) - \cdots$$
$$= \xi'_{A(-1)} \mathbf{q}_\alpha^d,$$

where $d = d(F, F_{A(-1)}) = \frac{1}{2}(d(A\sigma_F, A) - 1)$. Therefore

$$\xi_{A\sigma_F}^* = \mathbf{q}_\alpha^{\frac{d(A\sigma_F,A)-1}{2}} \xi_A^*.$$

The lemma is proved.

3.9. Denote ρ the half of the sum of all positive roots in R^+ and set $\rho^\vee = \frac{1}{2} \sum_{\alpha \in R^+} \alpha^\vee$.

Let W_0' be the subgroup of W_a generated by those simple reflections s such that the closure of sA_ρ^+ contains ρ. It is known that W_0' is a parabolic subgroup of W_a and is conjugate to W_0 under an element in Ω.

Let $u \in W_a$ be such that $u(A_0^+) = A_y^-$ for some y in $(X^+ + \rho) \cap (P + \rho)$. Note that $A_y^- \subset C_0^+$ since y is in $X^+ + \rho$. Consider the expression

$$T_u \sum_{w \in W_0} T_w A_0^- = T_u \sum_{\substack{A \in \Upsilon \\ \bar{A} \ni 0}} A = \sum_{\substack{A,B \in \Upsilon \\ \bar{A} \ni 0}} \pi^A_{B,A_y^-} B \in \mathcal{M}.$$

Since y is in $P + \rho$, we know that $B = T_w(A_x^-) = w(A_x^-)$ for certain x in $P + \rho$ and w in W_0' provided that π^A_{B,A_y^-} is not equal to zero. Note that for each z in W_0' one has

$$T_z \sum_{w \in W_0'} (-1)^{l(w)} \mathbf{q}_w T_w^{-1} = \sum_{w \in W_0'} (-1)^{l(w)} \mathbf{q}_w T_w^{-1} T_z$$

(3.9.1)
$$= (-1)^{l(z)} \sum_{w \in W_0'} (-1)^{l(w)} \mathbf{q}_w T_w^{-1}.$$

Therefore there exists a function $f : P + \rho \to \mathbb{Z}[\Gamma]$ such that

(3.9.2)
$$(\sum_{w \in W_0'} (-1)^{l(w)} \mathbf{q}_w T_w^{-1}) T_u (\sum_{w \in W_0} T_w) A_0^-$$
$$= \mathbf{q}_{w_0}^{-1} \sum_{w \in W_0'} (-1)^{l(w)} \mathbf{q}_w T_w^{-1} \sum_{x \in P + \rho} f(x) A_x^-.$$

Obviously we have

(a). The support of f is finite.

From Lemma 3.4 we see the following

(b). $f(y) = \mathbf{q}_{w_0}$, and $x \leq y$ if $f(x) \neq 0$. ($x \leq y$ means that there are non-negative integers n_α, $\alpha \in R^+$ such that $y - x = \sum_{\alpha \in R^+} n_\alpha \alpha$.)

Let α be a simple root and A be an alcove. Assume that $A = T_w(A_x^-)$ for some x in $P + \rho$ and w in W_0'. Let F be a hyperplane in $\mathcal{F}_\alpha$, then $A\sigma_F = T_{w'}(A_{x\sigma_F}^-)$ for some w' in W_0' such that $l(w') - l(w) \equiv 1 \pmod 2$. Now using (3.9.1) and applying Lemma 3.8 (ii) to $T_u(\sum_{w \in W_0} T_w) A_0^-$, we get the following

(c). Let α be a simple root and $z \in P + \rho$. Set $Y := \{z + n\alpha \mid n \in \mathbb{Z}\}$. Assume that a is a non-negative integer such that $\langle x, \alpha^\vee \rangle \equiv a \pmod 2$ for all x in Y. Then

$$\sum_{\substack{x \in Y \\ \langle x, \alpha^\vee \rangle \geq a}} f(x) = -\mathbf{q}_\alpha^{-(a-1)} \sum_{\substack{x \in Y \\ \langle x, \alpha^\vee \rangle \leq -a}} f(x).$$

If $u(A_0^+) = A_\rho^-$, then the function f can be given explicitly. To see it we need a few notations. For a subset I of R^+, set

$$\mathbf{q}_I := \prod_{\alpha \in I} \mathbf{q}_\alpha.$$

Let $x \in X$ and let $y, z \in X^+ := \{x \in X \mid \langle x, \alpha^\vee \rangle \geq 0 \text{ for all } \alpha \text{ in } R^+\}$ be such that $x = y - z$, we define

$$\mathbf{q}^x = \mathbf{q}_{p_y}^{\frac{1}{2}} \mathbf{q}_{p_z}^{-\frac{1}{2}}.$$

The element $\mathbf{q}^x$ is well defined since $\mathbf{q}_{p_{x'}}^{\frac{1}{2}} \mathbf{q}_{p_{x''}}^{\frac{1}{2}} = \mathbf{q}_{p_{x'+x''}}^{\frac{1}{2}}$ whenever x', x'' are in X^+. We have

(d). Let $x, y \in X$, then $\mathbf{q}^{x+y} = \mathbf{q}^x \mathbf{q}^y$.

(e). Let $\alpha_1, ..., \alpha_k$ be simple roots in R and let $\beta = a_1 \alpha_1 + \cdots + a_k \alpha_k \in (\frac{1}{2} P) \cap X$, then

$$\mathbf{q}^\beta = \mathbf{q}_{\alpha_1}^{a_1} \cdots \mathbf{q}_{\alpha_k}^{a_k}.$$

3.10. Lemma. *There exists a unique function* $f_0 : P + \rho \to \mathbb{Z}[\Gamma]$ *with a finite support satisfying properties (i), (ii), (iii) below:*

(i). $f_0(\rho) = \mathbf{q}_{w_0}$, *where* w_0 *is the longest element of* W_0.

33

(ii). $f_0(x) \neq 0$ implies that $x \leq \rho$.

(iii). Let $Y \subset P + \rho$ be an α-string: $Y = \{z + n\alpha \mid n \in \mathbb{Z}\}$, where z is any fixed element of $P + \rho$ and α is any fixed simple root. Let a be a non-negative integer such that $\langle x, \alpha^\vee \rangle = a \pmod 2$ for all x in Y. Then

$$\sum_{\substack{x \in Y \\ \langle x, \alpha^\vee \rangle \geq a}} f_0(x) = -\mathbf{q}_\alpha^{-(a-1)} \sum_{\substack{x \in Y \\ \langle x, \alpha^\vee \rangle \leq -a}} f_0(x).$$

The function is given by the formula

(3.10.1)
$$f_0(x) = (-1)^\nu \sum_{\substack{I \subseteq R^+ \\ \alpha_I = x + \rho}} (-1)^{|I|} \mathbf{q}_I \mathbf{q}^{\rho - x},$$

where α_I is the sum of all roots in I and $\nu = |R^+|$.

Proof. We first show that the function f_0 is unique, then prove that the function defined by (3.10.1) satisfies (i), (ii) and (iii).

Let g be a map from $P + \rho$ to $\mathbb{Z}[\Gamma]$ with a finite support. Assume that g satisfies the function equation in (iii) and $g(\rho) = 0$, $g(x) \neq 0 \Rightarrow x \leq \rho$. To see the uniqueness of the function f_0 it is enough to prove that $g(x) = 0$ for every x in $P + \rho$.

Assume that $g(y)$ is not equal to zero for some y in $P + \rho$. We may choose y such that (y, y) is maximal in $\{(x, x) \mid x \in P + \rho \text{ and } g(x) \text{ is not equal to zero}\}$. Let α be a simple root and let s be the corresponding simple reflection. Then $y' = (y)s$ is also in the string $Y := \{y + n\alpha \mid n \in \mathbb{Z}\}$. Let a be the absolute value of $(y, \alpha^\vee) = -(y', \alpha^\vee) = 2(y, \alpha)/(\alpha, \alpha)$. If $z = y + n\alpha \in Y$ satisfies $|(z, \alpha^\vee)| > a$, then $(z, z) = (y, y) + 2n(y, \alpha) + n^2(\alpha, \alpha) > (y, y)$. Hence $g(z) = 0$. Since g satisfies the function equation in (iii) , we get $g(y) = -\mathbf{q}_\alpha^{\pm(a-1)} g(y')$. It follows that $g(y')$ is not equal to zero. Note that $(y', y') = (y, y)$ since $(,)$ is W_0-invariant. Iterating the above process we see that $g((y)w)$ is not equal to zero for any w in W_0. Moreover $((y)w, (y)w) = (y, y)$.

For a suitable w in W_0 we have $\langle (y)w, \alpha^\vee \rangle \geq 0$ for each simple root α. Replacing y by $(y)w$, we may thus assume that $\langle y, \alpha^\vee \rangle \geq 0$ for all simple roots α. If $\langle y - \rho, \alpha^\vee \rangle \geq 0$ for all simple roots α, then $(y - \rho, \rho^\vee) \geq 0$. But $g(y) \neq 0$, so $\rho - y \geq 0$, hence $\rho - y = \sum n_\alpha \alpha$ (α simple roots, n_α non-negative integers). Therefore $(-\sum n_\alpha \alpha, \rho^\vee) \geq 0$. Thus $-\sum n_\alpha = 0$, hence $n_\alpha = 0$ for all simple roots α. So $y = \rho$. This leads to a contradiction between $g(y) \neq 0$ and $g(\rho) = 0$. Thus, there exists a simple root α such that $\langle y - \rho, \alpha^\vee \rangle < 0$. Since $\langle y, \alpha^\vee \rangle \geq 0$ and $\langle \rho, \alpha^\vee \rangle = 1$, we get $\langle y, \alpha^\vee \rangle = 0$. So $(y + n\alpha, y + n\alpha) > (y, y)$ if n is a nonzero integer. Thus $g(y + n\alpha) = 0$ for each nonzero integer n. Let us write down the function equation in (iii) for g, $Y := \{y + n\alpha \mid n \in \mathbb{Z}\}$ and $a = 0$. We get $g(y) = -\mathbf{q}_\alpha g(y)$, hence $g(y) = 0$. This contradiction shows that $g(x) = 0$ for every x in $P + \rho$.

Now we prove that the function f_0 defined by (3.10.1) satisfies (i), (ii) and (iii). The function f_0 defined by (3.10.1) clearly satisfies (i) and (ii). We now verify that it

34

satisfies (iii). In the following sums I and I' run through the set of all subsets of R^+. We have

$$\sum_{\substack{x\in Y \\ \langle x,\alpha^\vee\rangle\geq a}} f_0(x) = (-1)^\nu \sum_{\substack{x\in Y \\ I \\ \alpha_I=x+\rho \\ \langle x,\alpha^\vee\rangle\geq a}} (-1)^{|I|}\mathbf{q}_I\mathbf{q}^{\rho-x}$$

$$= (-1)^\nu \sum_{\substack{x\in Y \\ I\not\ni\alpha \\ \alpha_I=x+\rho \\ \langle x,\alpha^\vee\rangle\geq a}} (-1)^{|I|}\mathbf{q}_I\mathbf{q}^{\rho-x} + \Sigma',$$

where

$$\Sigma' = (-1)^\nu \sum_{\substack{x\in Y \\ I\ni\alpha \\ \alpha_I=x+\rho \\ \langle x,\alpha^\vee\rangle\geq a}} (-1)^{|I|}\mathbf{q}_I\mathbf{q}^{\rho-x} = (-1)^\nu \sum_{\substack{x\in Y \\ I'\not\ni\alpha \\ \alpha_{I'}=x-\alpha+\rho \\ \langle x,\alpha^\vee\rangle\geq a}} (-1)^{|I'|+1}\mathbf{q}_{I'}\mathbf{q}_\alpha\mathbf{q}^{\rho-x}$$

$$= (-1)^\nu \sum_{\substack{x'\in Y \\ I'\not\ni\alpha \\ \alpha_{I'}=x'+\rho \\ \langle x'+\alpha,\alpha^\vee\rangle\geq a}} (-1)^{|I'|+1}\mathbf{q}_{I'}\mathbf{q}_\alpha\mathbf{q}^{\rho-x'-\alpha} = -(-1)^\nu \sum_{\substack{x\in Y \\ I\ni\alpha \\ \alpha_I=x+\rho \\ \langle x,\alpha^\vee\rangle\geq a-2}} (-1)^{|I|}\mathbf{q}_I\mathbf{q}^{\rho-x}$$

Hence

(3.10.2)
$$\sum_{\substack{x\in Y \\ \langle x,\alpha^\vee\rangle\geq a}} f_0(x) = -(-1)^\nu \sum_{\substack{x\in Y \\ I\ni\alpha \\ \alpha_I=x+\rho \\ \langle x,\alpha^\vee\rangle=a-2}} (-1)^{|I|}\mathbf{q}_I\mathbf{q}^{\rho-x}.$$

A simialr computation shows that

$$\sum_{\substack{x\in Y \\ \langle x,\alpha^\vee\rangle\leq -a}} f_0(x) = (-1)^\nu \sum_{\substack{x\in Y \\ I\ni\alpha \\ \alpha_I=x+\rho \\ \langle x,\alpha^\vee\rangle=-a}} (-1)^{|I|}\mathbf{q}_I\mathbf{q}^{\rho-x}.$$

The simple reflection s corresponding to α maps $R^+ - \{\alpha\}$ onto $R^+ - \{\alpha\}$. Therefore the last sum is equal to

$$(-1)^\nu \sum_{\substack{x\in Y \\ I\ni\alpha \\ \alpha_{(I)_s}=(x+\rho)s \\ \langle x,\alpha^\vee\rangle=-a}} (-1)^{|I|}\mathbf{q}_I\mathbf{q}^{\rho-x} = (-1)^\nu \sum_{\substack{x\in Y \\ I'\not\ni\alpha \\ \alpha_{I'}=x+(a-1)\alpha+\rho \\ \langle x,\alpha^\vee\rangle=-a}} (-1)^{|I'|}\mathbf{q}_{I'}\mathbf{q}^{\rho-x}$$

$$= (-1)^\nu \sum_{\substack{x'\in Y \\ I'\not\ni\alpha \\ \alpha_{I'}=x'+\rho \\ \langle x'-(a-1)\alpha,\alpha^\vee\rangle=-a}} (-1)^{|I'|}\mathbf{q}_{I'}\mathbf{q}^{\rho+(a-1)\alpha-x'}$$

$$= (-1)^\nu \mathbf{q}_\alpha^{a-1} \sum_{\substack{x'\in Y \\ I'\not\ni\alpha \\ \alpha_{I'}=x'+\rho \\ \langle x',\alpha^\vee\rangle=a-2}} (-1)^{|I'|}\mathbf{q}_{I'}\mathbf{q}^{\rho-x'}.$$

35

Comparing with the right hand side of (3.10.2), we conclude that f_0 satisfies (iii). The lemma is proved.

Formula (3.9.2) and Lemma 3.10 imply the following

3.11. Lemma. *Let* $u \in W_a$ *be such that* $u(A_0^+) = A_\rho^-$, *then*

(3.11.1)
$$
(\sum_{w \in W_0'} (-1)^{l(w)} \mathbf{q}_w T_w^{-1}) T_u (\sum_{w \in W_0} T_w) A_0^-
$$
$$
= \mathbf{q}_{w_0}^{-1} (\sum_{w \in W_0'} (-1)^{l(w)} \mathbf{q}_w T_w^{-1}) \sum_{x \in P + \rho} f_0(x) h_x A_0^- ,
$$

where h_x is the unique element in H_φ' such that $h_x A_0^- = A_x^-$.

3.12. The equality (3.11.1) has some interesting consequences, which are essential to the proof of the main result Theorem 3.22. We first introduce some notations.

The set $X^+ := \{ x \in X \mid \langle x, \alpha^\vee \rangle \geq 0 \text{ for all } \alpha \text{ in } R^+ \}$ parametrizes the double cosets $W_0 \backslash W / W_0$, $x \leftrightarrow W_0 p_x W_0$. For x in X^+, denote m_x the shortest element in $W_0 p_x W_0$. Obviously $w_0 p_x$ is the longest element in $W_0 p_x W_0$ when x is in X^+. Let $W_x := \{ w \in W_0 \mid w p_x = p_x w \}$ be the stabilizer of x in W_0. Denote ν_x the length of the longest element w_x of W_x and set $\eta_x = \sum_{w \in W_x} \mathbf{q}_w$. When $x = 0$ we simply write ν, η for ν_0, η_0 respectively. Set

(3.12.1) $\quad K_x := \dfrac{1}{\eta} \displaystyle\sum_{w \in W_0 p_x W_0} T_w = \dfrac{\mathbf{q}_{w_0}^{-1} \mathbf{q}_{w_x}}{\eta \eta_x} (\sum_{w \in W_0} T_w) T_{p_x} (\sum_{w \in W_0} T_w), \quad x \in X^+,$

(3.12.2) $\quad J_x := (\displaystyle\sum_{w \in W_0} (-1)^{l(w)} \mathbf{q}_w T_w^{-1}) \mathbf{q}_{m_x}^{-\frac{1}{2}} T_{m_x} (\sum_{w \in W_0} T_w).$

For each s in S_0 we have

$$
T_s (\sum_{w \in W_0} T_w) = (\sum_{w \in W_0} T_w) T_s = \mathbf{q}_s (\sum_{w \in W_0} T_w),
$$

and

$$
\sum_{w \in W_0} (-1)^{l(w)} \mathbf{q}_w T_w^{-1} = h_s (1 - \mathbf{q}_s T_s^{-1}) \quad \text{for some } h_s \text{ in } H_\varphi.
$$

Since $m_x = p_x w_x$, we obtain

$$
J_x = 0 \quad \text{for } x \text{ in } X^+ - (X^+ + \rho),
$$

(3.12.3) $\quad J_x = \mathbf{q}_{w_0}^{-\frac{1}{2}} (\displaystyle\sum_{w \in W_0} (-1)^{l(w)} \mathbf{q}_w T_w^{-1}) \mathbf{q}_{p_x}^{-\frac{1}{2}} T_{p_x} (\sum_{w \in W_0} T_w) \text{ for } x \in X^+ + \rho.$

36

Then $K_x, x \in X^+$ form a $\mathbb{Z}[\Gamma]$-basis of

$$\mathcal{K} := \{ h \in \frac{1}{\eta} H_\varphi \mid (\sum_{w \in W_0} T_w) h = h(\sum_{w \in W_0} T_w) = \eta x \} \subset H_\varphi \otimes \mathbb{Q}(\Gamma),$$

and $J_x, x \in X^+ + \rho$ form a $\mathbb{Z}[\Gamma]$-basis of

$$\mathcal{J} := \{ h \in H_\varphi \mid (\sum_{w \in W_0} (-1)^{l(w)} \mathbf{q}_w T_w^{-1}) h = h(\sum_{w \in W_0} T_w) = \eta h \}.$$

Note that $\mathcal{K}$ is a subring of $H_\varphi \otimes \mathbb{Q}(\Gamma)$ with unit $\frac{1}{\eta}(\sum_{w \in W_0} T_w)$, and with respect to the multiplication in $H_\varphi \otimes \mathbb{Q}(\Gamma)$ we have $\mathcal{J} \cdot \mathcal{K} \subseteq \mathcal{J}$, that is, $\mathcal{J}$ is a right $\mathcal{K}$-module. (Where $\mathbb{Q}(\Gamma)$ is the quotient field of $\mathbb{Z}[\Gamma]$.)

To be convenient we define J_x for any x in X as follows. If $(x)w \neq x$ for all $w \in W_0$, $w \neq e$, we set $J_x = (-1)^{l(w)} J_{(x)w}$, where w is the unique element of W_0 such that $(x)w \in X^+ + \rho$. For the remaining x in X we set $J_x = 0$. We also write C for $\sum_{w \in W_0} T_w$ and write C' for $\sum_{w \in W_0} (-1)^{l(w)} \mathbf{q}_w T_w^{-1}$.

3.13. Let $x \in X$ and let s be a simple reflection in W_0, then we have (cf. 2.13 for the naotation θ_x)

$$(3.13.1) \qquad T_s(\theta_x + \theta_{(x)s}) = (\theta_x + \theta_{(x)s}) T_s.$$

For each $x \in X$, set $\tilde{J}_x := \mathbf{q}_{w_0}^{-\frac{1}{2}} C' \theta_x C \in J$. When x is in $X^+ + \rho$, clearly we have $\tilde{J}_x = J_x$. In general we have

(a). $\tilde{J}_{(x)w} = (-1)^{l(w)} \tilde{J}_x$ for any x in X and w in W_0. Hence $\tilde{J}_x = J_x$ for all x in X.

Proof. We may assume that $w = s$ is a simple reflection in W_0. Since $T_s C = \mathbf{q}_s C$ and $C' T_s^{-1} = -C'$, we see

$$\tilde{J}_x + \tilde{J}_{(x)s} = \mathbf{q}_{w_0}^{-\frac{1}{2}} C'(\theta_x + \theta_{(x)s}) C$$

$$= \mathbf{q}_{w_0}^{-\frac{1}{2}} C' T_s^{-1}(\theta_x + \theta_{(x)s}) T_s C$$

$$= -\mathbf{q}_s \mathbf{q}_{w_0}^{-\frac{1}{2}} C'(\theta_x + \theta_{(x)s}) C$$

$$= -\mathbf{q}_s(\tilde{J}_x + \tilde{J}_{(x)s}).$$

Hence $\tilde{J}_x + \tilde{J}_{(x)s} = 0$, as required.

3.14. Lemma. For every x in X^+ we have

$$(3.14.1) \qquad J_\rho(\mathbf{q}_{p_x}^{-\frac{1}{2}} K_x) = \frac{1}{\bar{\eta}_x} \sum_{I \subseteq R^+} (-1)^{|I|} \mathbf{q}_I J_{x+\rho-\alpha_I}.$$

(See 3.9 for notation.)

37

Proof. Since the H'_φ-module $\mathcal{M}$ is faithful, we can remove A_0^- from the two sides of (3.11.1) and obtain an identity in H'_φ:

(3.14.2)
$$\left(\sum_{w\in W'_0} (-1)^{l(w)}\mathbf{q}_w T_w^{-1}\right)T_u\left(\sum_{w\in W_0} T_w\right) = \mathbf{q}_{w_0}^{-1}\sum_{w\in W'_0}(-1)^{l(w)}\mathbf{q}_w T_w^{-1}\sum_{x\in P+\rho}f_0(x)h_x.$$

Let $\omega \in \Omega$ be such that $\omega W'_0\omega^{-1} = W_0$. Then $T_\omega T_u = T_{\omega u} = T_{m_\rho}$ and $T_\omega h_x = \mathbf{q}^x\theta_x$ (see 3.9 for definition). Multiply both sides of identity (3.14.1) on the left by T_ω we get

(3.14.3)
$$\left(\sum_{w\in W_0} (-1)^{l(w)}\mathbf{q}_w T_w^{-1}\right)T_{m_\rho}\left(\sum_{w\in W_0} T_w\right) = \mathbf{q}_{w_0}^{-1}\sum_{w\in W_0}(-1)^{l(w)}\mathbf{q}_w T_w^{-1}\sum_{x\in P+\rho}f_0(x)\mathbf{q}^x\theta_x.$$

Now we have

$$\begin{aligned}
J_\rho(\mathbf{q}_{p_x}^{-\frac{1}{2}}K_x) &= \mathbf{q}_{m_\rho}^{-\frac{1}{2}}C'T_{m_\rho}C\frac{\mathbf{q}_{w_0}^{-1}\mathbf{q}_{w_x}}{\eta\eta_x}C\mathbf{q}_{p_x}^{-\frac{1}{2}}T_{p_x}C \\
&= \frac{1}{\eta_x}\mathbf{q}_{m_\rho}^{-\frac{1}{2}}\mathbf{q}_{w_0}^{-2}\mathbf{q}_{w_x}\sum_{y\in P+\rho}f_0(y)\mathbf{q}^y C'\theta_x\theta_y C \\
&= \frac{1}{\bar\eta_x}\mathbf{q}_{p_\rho}^{-\frac{1}{2}}\mathbf{q}_{w_0}^{\frac{1}{2}}\mathbf{q}_{w_0}^{-2}(-1)^\nu\sum_{I\subseteq R^+}(-1)^{|I|}\mathbf{q}_I\mathbf{q}_{p_\rho}^{\frac{1}{2}}\mathbf{q}_{w_0}^{\frac{1}{2}}J_{x+\alpha_I-\rho} \quad \text{(by 3.13 (a))}\\
&= \frac{1}{\bar\eta_x}\sum_{I\subseteq R^+}(-1)^{\nu-|I|}\mathbf{q}_I\mathbf{q}_{w_0}^{-1}J_{x+\alpha_I-\rho}.
\end{aligned}$$

Let $I' := R^+ - I$ be the complement of I in R^+, then $\alpha_I + \alpha_{I'} = 2\rho$. Hence we have

$$\frac{1}{\bar\eta_x}\sum_{I\subseteq R^+}(-1)^{\nu-|I|}\mathbf{q}_I\mathbf{q}_{w_0}^{-1}J_{x+\alpha_I-\rho} = \frac{1}{\bar\eta_x}\sum_{I'\subseteq R^+}(-1)^{|I'|}\mathbf{q}_{I'}^{-1}J_{x+\rho-\alpha_{I'}}.$$

The lemma is proved.

3.15. Theorem. *For each x in X^+ there is a unique element $\tilde{C}_x$ in $\mathcal{K}$ satisfies the following conditions:*

(i). $J_\rho\tilde{C}_x = J_{x+\rho}.$

(ii). $\tilde{C}_x = \mathbf{q}_{p_x}^{-\frac{1}{2}}\sum_{\substack{y\in X^+ \\ y\le x}} d_{y,x}K_y.$

(iii). *Each $d_{y,x}$ is a Laurant polynomial in $\mathbf{q}_s$, $s \in S$ with integer coefficients. Moreover $\mathbf{q}_{p_x}^{-\frac{1}{2}}\mathbf{q}_{p_y}^{\frac{1}{2}}d_{y,x}$ is a polynomial in $\mathbf{q}_s^{-1}$, $s\in S$ without constant term if $y < x$ and $d_{x,x} = 1$.*

In particular, the map $h \to J_\rho h$ is an isomorphism of right $\mathcal{K}$-module from $\mathcal{K}$ to $\mathcal{J}$.

(Compare with [L5, Corollary 6.8].)

Proof. Let I be a subset of R^+. Assume that there is $w \in W_0$ such that $x + \rho - \alpha_I = (x' + \rho)w$ for some x' in X^+. Then $x - x' = x - (x)w^{-1} - (\rho)w^{-1} + \rho + (\alpha_I)w^{-1} = x - (x)w^{-1} + \alpha_J$, where J is the set of positive roots β such that $(\beta)w \in I$ or such that $-(\beta)w \in R^+ - I$. Since $x \geq (x)w^{-1}$ and $\alpha_J \geq 0$, we have $x \geq x'$. Thus the right hand side of (3.14.1) is a linear combination of elements $J_{x'+\rho}$ ($x' \leq x$) with formal power series in $\mathbf{q}_s^{-1}$ and each term has no factors $\mathbf{q}_s^i$, $i \geq 1$, $s \in S$. Moreover the coefficient doesnot have constant term if $x' < x$. However, the left hand side of (3.14.1) is in $\mathcal{J}$, so these coefficients must be polynomials in $\mathbf{q}_s^{\frac{1}{2}}, \mathbf{q}_s^{-\frac{1}{2}}$. Hence they are polynomials in $\mathbf{q}_s^{-1}, s \in S$ (without constant term if $x' < x$).

We claim that the coefficient of $J_{x+\rho}$ is 1. Actually, let I be a subset of R^+ and w be an element in W_x such that $\alpha_I = \rho - (\rho)w$. By the following Lemma 3.16, $I = \{\alpha \in R^+ \mid (\alpha)w^{-1} < 0\}$, hence $\mathbf{q}_w = \mathbf{q}_I$. Now the claim follows from the identity $\dfrac{1}{\eta_x} \displaystyle\sum_{\substack{I \subseteq R^+ \\ w \in W_x \\ \alpha_I = \rho - (\rho)w}} \mathbf{q}_I^{-1} = 1$.

A triangular matrix with 1 on the diagonal entries has an inverse of the same form, so for each x in X^+ the element $J_{x+\rho}$ is a linear combination of the elements $J_\rho(\mathbf{q}_{p_y}^{-\frac{1}{2}} K_y)$, $y \leq x$, whose coefficients are polynomials in $\mathbf{q}_s^{-1}, s \in S$ (without constant term if $y < x$ and equal to 1 if $y = x$). That is

$$J_{x+\rho} = J_\rho \sum_{\substack{y \in X^+ \\ y \leq x}} \xi_{y,x} \mathbf{q}_{p_y}^{-\frac{1}{2}} K_y,$$

where $\xi_{y,x}$ is a polynomial in $\mathbf{q}_s^{-1}$, $s \in S$ (without constant term if $y < x$ and equal to 1 if $y = x$. Let

$$d_{y,x} = \mathbf{q}_{p_x}^{\frac{1}{2}} \mathbf{q}_{p_y}^{-\frac{1}{2}} \xi_{y,x}.$$

Since $y \leq x$, we can find integers $n_1, ..., n_k$ and simple roots $\alpha_1, ..., \alpha_k$ such that $x - y = n_1 \alpha_1 + \cdots + n_k \alpha_k$. Then $d_{y,x} = (\prod_{i=1}^{k} \mathbf{q}_{\alpha_i}^{n_i}) \xi_{y,x}$ is a Laurant polynomials in $\mathbf{q}_s, s \in S$. Moreover we have $\mathbf{q}_{p_x}^{-\frac{1}{2}} \mathbf{q}_{p_y}^{\frac{1}{2}} d_{y,x} = \xi_{y,x} \in \mathbb{Z}[\mathbf{q}_s^{-1}]_{s \in S}$, which has no constant term if $y < x$ and is equal to 1 if $y = x$.

The theorem is proved.

3.16. Lemma. *For every $w \in W_0$ there is a unique subset I of R^+ such that $\alpha_I = \rho - (\rho)w$ and the unique subset is $\{\alpha \in R^+ \mid (\alpha)w^{-1} < 0\}$.* [Ko, 5.10.2].

Proof. Let $I := \{\alpha \in R^+ \mid (\alpha)w^{-1} < 0\}$. Then obviously we have $\alpha_I = \rho - (\rho)w$. We use induction on the length of w to prove the uniqueness. When $l(w) = 0, 1$, the uniqueness is obvious. Now assume that $l(w) \geq 2$ and let s be a simple reflection such that $u = ws \leq w$. Let β be the simple root corresponding to s and J be a subset of R^+ such that $\alpha_J = \rho - (\rho)w$. If β is not in J, then $Js \subseteq R^+ - \{\beta\}$ and $\alpha_{J'} = \rho - (\rho)u$, where $J' = Js \cup \{\beta\}$. By induction hypothesis we have $J' = \{\alpha \in R^+ \mid (\alpha)u^{-1} < 0\}$. This is impossible since $(\beta)u^{-1} > 0$. Hence β is in J and $\alpha_K = \rho - (\rho)u$, where $K = Js - \{\beta\}$. By induction hypothesis we have $K = \{\alpha \in R^+ \mid (\alpha)u^{-1} < 0\}$. Therefore $J = Ks \cup \{\beta\} = I$. The lemma is proved.

To see the relation between $\tilde{C}_x$ and $C_{w_0 p_x}$ we need the following result.

3.17. Lemma. *If $x \in X^+$, then $\bar{J}_{x+\rho} = J_{x+\rho}$.*

Proof. Let

$$C = \sum_{w \in W_0} T_w, \qquad C' = \sum_{w \in W_0} (-1)^{l(w)} \mathbf{q}_w T_w^{-1}.$$

Then

$$\overline{\mathbf{q}_{w_0}^{-\frac{1}{2}} C} = \mathbf{q}_{w_0}^{-\frac{1}{2}} C, \qquad \overline{\mathbf{q}_{w_0}^{\frac{1}{2}} C'} = \mathbf{q}_{w_0}^{\frac{1}{2}} C'.$$

So it is enough to prove the following equality for all z in X^+,

(3.17.1) $$C'(\mathbf{q}_{m_z} T_{m_z^{-1}}^{-1} - T_{m_z}) C = 0.$$

Note that $m_z = p_z w_z$ and $T_{p_z} = T_{m_z} T_{w_z}$. Since $T_{w_z} C = \mathbf{q}_{w_z} C$ and for every s in S_0 there is some h_s in H_φ such that $C' = h_s(1 - \mathbf{q}_s T_s^{-1})$, so

$$C' T_{m_z^{-1}}^{-1} C = \mathbf{q}_{w_z} C' T_{p_z}^{-1} C = 0, \qquad C' T_{m_z} C = \mathbf{q}_{w_z}^{-1} C' T_{p_z} C = 0,$$

provided that z is not in $X^+ + \rho$.

Now we assume that $z \in X^+ + \rho$. For $w = w' p_x, w' \in W_0, x \in X$, denote w^* the element $w_0(w' p_x^{-1}) w_0^{-1}$. The map $w \to w^*$ then defines an involution of W which sends S to S. Consider two $\mathbb{Z}[\Gamma]$-linear maps Ψ, $\tilde{\Psi} : H_\varphi \to H_\varphi$ defined by $\Psi(T_w) = (-1)^{l(w)} \mathbf{q}_w T_w^{-1}$ and $\tilde{\Psi}(T_w) = \Psi(T_{w^*})$. Both maps are antiautomorphism of order 2. It is clear that $\tilde{\Psi}(C) = C'$, $\tilde{\Psi}(C') = C$. We have $m_z^{-1} = m_z^*$. So

$$\tilde{\Psi}(C' T_{m_z} C) = \tilde{\Psi}(C)(-1)^{l(m_z)} \mathbf{q}_{m_z} T_{m_z^*}^{-1} \tilde{\Psi}(C') = C'(-1)^{l(m_z)} \mathbf{q}_{m_z} T_{m_z^{-1}}^{-1} C$$

$$= (-1)^{l(m_z)} C'(T_{m_z} + \sum_{y<z} f_{y,z} T_{m_y}) C \qquad (f_{y,z} \in \mathbb{Z}[\Gamma])$$

$$= (-1)^{l(m_z)} C' T_{m_z} C + (\text{combination of elements } C' T_{m_y} C),$$

where $y \in X^+ + \rho$ and $l(m_y) < l(m_z)$.

Thus the matrix of $\tilde{\Psi}$ with respect to the basis elements $C' T_{m_z} C$ ($z \in X^+ + \rho$) is upper triangular with ± 1 on the diagonal. On the other hand, it must have order 2, hence non-diagonal entries of the matrix must be zero. Thus $(-1)^{l(m_z)} C' T_{m_z} C = (-1)^{l(m_z)} C'(\mathbf{q}_{m_z} T_{z^{-1}}^{-1}) C$. This is just what we need.

3.18. Theorem. *Assume that we have a total order on Γ which is compatible with the multiplication of Γ and all $\mathbf{q}_s^{\frac{1}{2}}$ are positive. Let $x \in X^+$, then*

$$\tilde{C}_x = \mathbf{q}_{w_0}^{\frac{1}{2}} \eta^{-1} C_{w_0 p_x}.$$

Proof. Extend the involution $\bar{}$ of H_φ to $H_\varphi \otimes \mathbb{Q}(\Gamma)$ in an obvious way. and note that $\overline{\eta^{-1} C} = \eta^{-1} C$. The definition of $\mathcal{K}$ shows that $\mathcal{K}$ is stable under the involution $\bar{}$.

From Theorem 3.15 (i) we get $J_\rho \tilde{\bar{C}}_x = J_{x+\rho}$. Thus $J_\rho(\tilde{C}_x - \tilde{\bar{C}}_x) = 0$. According to Theorem 3.15 (iii), it implies that $\tilde{C}_x = \tilde{\bar{C}}_x$ since $\tilde{C}_x - \tilde{\bar{C}}_x$ is in $\mathcal{K}$. The element $\mathbf{q}_{w_0}^{-\frac{1}{2}} \eta \tilde{C}_x$ is also fixed by the involution $\bar{}$ since $\overline{\mathbf{q}_{w_0}^{-\frac{1}{2}} \eta} = \mathbf{q}_{w_0}^{-\frac{1}{2}} \eta$. We have

$$\mathbf{q}_{w_0}^{-\frac{1}{2}} \eta \tilde{C}_x = \mathbf{q}_{w_0 p_x}^{-\frac{1}{2}} \sum_{\substack{y \in W \\ y \leq w_0 p_x}} d_{\mu(y),x} T_y,$$

where $\mu(y) \in X^+$ is defined by $y \in W_0 p_{\mu(y)} W_0$. From Theorem 3.15 (iii) we see that $\mathbf{q}_{w_0}^{-\frac{1}{2}} \eta \tilde{C}_x$ has the required properties for $C_{w_0 p_x}$ (cf. 1.14). The theorem is proved.

3.19. Recall that we have defined the element $U_x \in H_\varphi$ for every $x \in X^+$ (see 2.13). We have

(a). $J_\rho U_x = J_{x+\rho}$ for each $x \in X^+$.

Proof. By Weyl's character formula we get

$$\left(\sum_{w \in W_0} (-1)^{l(w)} \theta_{(\rho)w} \right) U_x = \sum_{w \in W_0} (-1)^{l(w)} \theta_{(x+\rho)w}.$$

Therefore we have

$$J_\rho U_x = |W_0|^{-1} \sum_{w \in W_0} (-1)^{l(w)} J_{(\rho)w} U_x$$

$$= |W_0|^{-1} \sum_{w \in W_0} \mathbf{q}_{w_0}^{-\frac{1}{2}} (-1)^{l(w)} C' \theta_{(\rho)w} C U_x \quad \text{using 3.13 (a)}$$

$$= |W_0|^{-1} C' \sum_{w \in W_0} \mathbf{q}_{w_0}^{-\frac{1}{2}} (-1)^{l(w)} \theta_{(\rho)w} U_x C$$

$$= |W_0|^{-1} C' \sum_{w \in W_0} \mathbf{q}_{w_0}^{-\frac{1}{2}} (-1)^{l(w)} \theta_{(x+\rho)w} C$$

$$= |W_0|^{-1} \sum_{w \in W_0} (-1)^{l(w)} J_{(x+\rho)w}$$

$$= J_{x+\rho}.$$

3.20. Theorem. (i). *The map $\Phi_1 : \mathcal{Z} \to \mathcal{J}$, $z \to J_\rho z$ defines an isomorphism of $\mathbb{Z}[\Gamma]$-algebra from the center $\mathcal{Z} := \mathcal{Z}(H_\varphi)$ of H_φ to $\mathcal{J}$.*

(ii). *The map $\Phi : \mathcal{Z} \to \mathcal{K}$ defined by $z \to \dfrac{1}{\eta} \sum_{w \in W_0} T_w z = \eta^{-1} C z$ is an isomorphism of $\mathbb{Z}[\Gamma]$-algebra preserving the unit element. Under this isomorphism each U_x ($x \in X^+$) is corresponding to $\tilde{C}_x \in \mathcal{K}$, that is $\tilde{C}_x = \eta^{-1} C U_x$.*

Proof. Part (i) follows from 3.19 (a). Now we prove (ii).

Denote Φ_2 the map $\mathcal{K} \to \mathcal{J}$, $K \to J_\rho K$. Since $\eta^{-1} J_\rho C = J_\rho$, we have $\Phi_1 = \Phi_2 \Phi$. Since both Φ_1 and Φ_2 are bijective, so Φ is bijective. Obviously Φ preserves the unit.

Moreover Φ preserves multiplication: $\Phi(z)\Phi(z') = \eta^{-1}Cz\eta^{-1}Cz' = \eta^{-2}C^2zz' = \eta^{-1}Czz' = \Phi(zz')$. Finally for x in X^+ we have

$$\Phi(U_x) = \Phi_2^{-1}\Phi_1(U_x) = \Phi_2^{-1}(J_{x+\rho}) = \tilde{C}_x.$$

The theorem is proved.

3.21. Theorem. *Let* x, y *be in* X^+. *Then*

(i). $C_{w_0p_x} = \mathbf{q}_{w_0}^{-\frac{1}{2}}CU_x = \mathbf{q}_{w_0}^{-\frac{1}{2}}(\sum_{w\in W_0} T_w)U_x.$

(ii). $C_{w_0p_x}C_{w_0p_y} = \sum_{z\in X^+} m(x,y,z)\mathbf{q}_{w_0}^{-\frac{1}{2}}\eta C_{w_0p_z}$, *where* $m(x,y,z)$ *is defined by* $U_xU_y = \sum_{z\in X^+} m(x,y,z)U_z$.

Proof. Part (i) follows from Theorem 3.20 and Theorem 3.18. Part (ii) follows from Theorem (i), also follows from 3.20 (ii).

Now we can prove the main result of the chapter.

3.22. Theorem. *Assume that we have a total order on* Γ *which is compatible with the multiplication in* Γ *and all* $\mathbf{q}_s^{\frac{1}{2}}$, $s \in S$ *are positive. Then the set*

$$c_0 = \{ww_0u \in W \mid l(ww_0u) = l(w) + l(w_0) + l(u)\}$$

is a generalized two-sided cell of W *with respect to the total order and* φ.

Proof. Set

$$\mathcal{L} := \{w \in W \mid sw \le w \text{ for each } s \text{ in } S_0\},$$
$$\mathcal{R} := \{w \in W \mid ws \le w \text{ for each } s \text{ in } S_0\}.$$

By (1.14.8-11) we get

(a). Let $w \in W$. Assume that $C_wC_s = (\mathbf{q}_s^{\frac{1}{2}} + \mathbf{q}_s^{-\frac{1}{2}})C_w$ for every simple reflection s in W_0, then $w \in \mathcal{R}$. If $C_sC_w = (\mathbf{q}_s^{\frac{1}{2}} + \mathbf{q}_s^{-\frac{1}{2}})C_w$ for every simple reflection s in W_0, then $w \in \mathcal{L}$.

Therefore we have (see 1.19)

(b). The elements $C_w, w \in \mathcal{R}$ form a $\mathbb{Z}[\Gamma]$-basis of $H_\varphi C_{w_0}$ and the elements $C_w, w \in \mathcal{L}$ form a $\mathbb{Z}[\Gamma]$-basis of $C_{w_0}H_\varphi$.

Using induction on $l(u)$ we see that

(c). For any $h \in H_\varphi$, the element $hC_{w_0}C_u$ is a $\mathbb{Z}[\Gamma]$-linear combination of $C_w, w \in c_0$.

Hence we have

(d). The elements $C_w, w \in c_0$ form a $\mathbb{Z}[\Gamma]$-basis of $H_\varphi C_{w_0}H_\varphi$.

Let $w \in c_0$. By the definition in 1.15 we clearly have $w \underset{L\overline{R},\varphi}{\le} w_0$. Let $x \in X^+$ be such that $w \in W_0p_xW_0$. Then we can find elements w', w'' in W_0 such that $w_0p_x = w'ww''$ and $l(w_0p_x) = l(w') + l(w) + l(w'')$. Thus we get $w_0p_x \underset{L\overline{R},\varphi}{\le} w$. Denote $(-x)w_0$ by x^*, then $m(x, x^*, 0) = 1$. By Theorem 3.21 (ii) and 1.15 (a) we

42

see $w_0 \underset{L\bar{R},\varphi}{\leq} w_0 p_x$. Therefore $w \underset{L\bar{R},\varphi}{\sim} w_0$. By (d) we know that if $u \underset{L\bar{R},\varphi}{\leq} w_0$, then $u \in c_0$, so $u \underset{L\bar{R},\varphi}{\sim} w_0$.

The theorem is proved.

From the proof we also see that $\mathcal{R}$ is a generalized left cell of W with respect to the given order and φ. Finally we conclude this chapter with two questions.

3.23. Question. (i). *How many generalized left cells are contained in c_0.*

(ii). *Whether the number of generalized two-sided cells of W is finite. Further whether the number of generalized left cells of W is finite.*

4. q$_s$-Analogue of Weight Multiplicity

In this chapter we will show that the main results in [Ka2] hold in a slightly more general form. We follow the approach of Gupta [Gu] to prove it. Theorem 4.5 is helpful to understand the elements $\tilde{C}_x$ defined in Theorem 3.15. The results in [Br] should hold also in a slightly more general form, but the author has no idea how to prove it (see 4.8).

4.1. Let V be an Euclidean space and let $R \subset V$ be an irreducible root system of rank dim$V$. We assume that the scalar product (,) of V is invariant under the Weyl group W of R. We shall write $\alpha^\vee$ rather than $2\alpha/(\alpha, \alpha)$ for each root α in R. For $v \in V$ and $\alpha \in R$ we also write $\langle v, \alpha^\vee \rangle$ instead of $(v, \alpha^\vee)$.

The set $X := \{x \in V \mid \langle x, \alpha^\vee \rangle$ is an integer for each α in $R\}$ is W-invariant and the root lattice $\mathbb{Z}R$ is a W-invariant subset of X. We shall denote e^λ the element $\lambda \in X$ when it is regarded as an element in the group algebra $\mathbb{Z}[X]$. The operation in X will be written additively. Then $e^{\lambda+\mu} = e^\lambda e^\mu$.

Let R^+ the set of positive roots in R and Δ the set of simple roots in R. Denote S the set of simple reflections in W. Let $\mathbf{g}$ be a simple Lie algebra over $\mathbb{C}$ of root system R. For each element λ in $X^+ = \{\mu \in X \mid \langle \mu, \alpha^\vee \rangle \geq 0$ for all α in $R^+\}$ we denote $\chi_\lambda \in \mathbb{Z}[X]$ the character of the irreducible $\mathbf{g}$-module of highest weight λ. Denote $\varepsilon(w)$ for $(-1)^{l(w)}$, and let ρ stand for the half of the sum of all positive roots in R^+. Then Weyl's character formula says

$$\chi_\lambda = \sum_{w \in W} \varepsilon(w)e^{w(\lambda+\rho)} \Big/ \sum_{w \in W} \varepsilon(w)e^{w(\rho)}.$$

For any $\lambda \in X$, we set $\chi_\lambda := \varepsilon(w)\chi_{w(\lambda+\rho)-\rho}$ if there exists some w in W such that $w(\lambda + \rho) - \rho \in X^+$ and set $\chi_\lambda = 0$ otherwise. The map $\lambda \to \chi_\lambda$ defines a homomorphism (of abelian group) χ from $\mathbb{Z}[X]$ to $\mathbb{Z}[X]^W$. We also write $\chi(\sum a_\lambda \lambda)$ for $\sum a_\lambda \chi_\lambda$. We define a scalar product on $\mathbb{Z}[X]^W$ by $\langle \chi_\lambda, \chi_\mu \rangle = \delta_{\lambda,\mu}$, $\lambda, \mu \in X^+$.

We are interested in some q$_s$-analogue of Weyl's character formula and of weight multiplicity. Recall that S is the set of simple reflections in W.

4.2. Let $\mathbf{q}_s$, $s \in S$ be indeterminates. We assume that $\mathbf{q}_s = \mathbf{q}_t$ if and only if s, t are conjugate in W. Let α be a simple root and s the corresponding simple reflection. We set $\mathbf{q}_\beta = \mathbf{q}_\alpha = \mathbf{q}_s$ provided that β is a root conjugate to α. For any reduced expression $w = s_1 s_2 \cdots s_k$ in W we set $\mathbf{q}_w = \mathbf{q}_{s_1}\mathbf{q}_{s_2} \cdots \mathbf{q}_{s_k}$. The element $\mathbf{q}_w$ is well defined because of the assumption on $\mathbf{q}_s$, $s \in S$.

Given an element $\lambda \in X$ we denote W_λ the stabilizer of λ in W. We also set $\eta_\lambda = \sum_{w \in W_\lambda} \mathbf{q}_w$.

Define

$$\Phi := \prod_{\alpha \in R^+} (1 - \mathbf{q}_\alpha e^{-\alpha}), \qquad \Psi := \prod_{\alpha \in R^+} (1 - \mathbf{q}_\alpha e^\alpha)^{-1}.$$

44

We regard them as elements in the ring Γ of power series in $\mathbf{q}_s, s \in S$ with coefficients in $\mathbb{Z}[X]$. We shall denote again $e^\lambda \in \Gamma$ the element corresponding to $\lambda \in X$. The map $\chi : \mathbb{Z}[X] \to \mathbb{Z}[X]^W$ may be extended linearly to a map $\Gamma \to \Gamma^W$, and the scalar product $\langle,\rangle$ can be extended linearly to Γ^W. Denote the extentions again by χ and $\langle,\rangle$ respectively.

Further, for $\lambda, \mu \in X^+$ we define

$$P_\lambda := \chi(e^\lambda \Phi)\eta_\lambda^{-1}, \qquad Q_\mu := \chi(e^\mu \Psi).$$

Then P_λ and Q_μ are elemenets in Γ^W.

The map $\chi_\lambda \to \chi_\lambda^* = \chi_{-w_0(\lambda)}$, $\lambda \in X$ defines an involution $*$ of Γ^W. The following simple facts will be needed in the proof of Theorem 4.3.

(a). If $\xi \in \Gamma$ is W-invariant, then $\chi(f\xi) = \chi(f)\xi$ for all f in Γ.

(b). Let ξ be an element of Γ^W, then $\langle f, g\xi \rangle = \langle f\xi^*, g \rangle$ for all $f, g \in \Gamma^W$.

4.3. Theorem. *For arbitrary* λ, μ *in* X^+ *we have* $\langle P_\lambda, Q_\mu \rangle = \delta_{\lambda,\mu}$. *(Compare with [Gu, Theorem 2.5].)*

Proof. We have

$$\Phi = \sum_{I \subseteq R^+} (-1)^{|I|} \mathbf{q}_I e^{-\alpha_I},$$

where $|I|$ denotes the cardinal of I, $\mathbf{q}_I = \prod_{\alpha \in I} \mathbf{q}_\alpha$, and $\alpha_I = \sum_{\alpha \in I} \alpha$. Therefore we have

$$P_\lambda = \eta_\lambda^{-1} \sum_{I \subseteq R^+} (-1)^{|I|} \mathbf{q}_I \chi_{\lambda - \alpha_I} = \sum_{\nu \in X^+} f_\nu \chi_\nu.$$

According to the proof of Theorem 3.15 we get

(a). If f_ν is not zero, then $\lambda - 2\rho \leq \nu \leq \lambda$. Moreover $f_\lambda = 1$.

Denote $\mathbb{N}^{R^+}$ the set of all maps from R^+ to $\mathbb{N}$. For each $\varphi : R^+ \to \mathbb{N}$, we write $\mathbf{q}_\varphi$ for $\prod_{\alpha \in R^+} \mathbf{q}_\alpha^{\varphi(\alpha)}$, and use α_φ for $\sum_{\alpha \in R^+} \varphi(\alpha)\alpha$. Then

$$(4.3.1) \qquad \Psi = \sum_{\varphi \in \mathbb{N}^{R^+}} \mathbf{q}_\varphi e^{\alpha_\varphi},$$

and

$$Q_\mu = \sum_{\varphi \in \mathbb{N}^{R^+}} \mathbf{q}_\varphi \chi_{\mu + \alpha_\varphi} = \sum_{\nu \in X^+} g_\nu \chi_\nu.$$

Let φ be an element in $\mathbb{N}^{R^+}$ and assume that $\nu + \rho = w(\mu + \alpha_\varphi + \rho)$ for some ν in X^+. Then $\nu + \rho \geq w^{-1}(\nu + \rho) \geq \mu + \rho$, hence $\nu \geq \mu$. Moreover $\nu = \mu$ if and only if w is the neutral element of W and $\varphi(\alpha) = 0$ for all α in R^+. Thus we obtain

(b). Assume that g_ν is not zero, then $\nu \geq \mu$. Moreover $g_\mu = 1$.

Set $\xi = \prod_{\alpha \in R}(1 - \mathbf{q}_\alpha e^\alpha)^{-1}$. Then we have $\Psi = \Phi\xi$. Since ξ is W-invariant and $\xi^* = \xi$, we obtain (cf. 4.2 (a-b))

(c). $\qquad \langle P_\lambda, Q_\mu \rangle = \langle P_\lambda, P_\mu\xi \rangle = \langle P_\lambda\xi^*, P_\mu \rangle = \langle Q_\lambda, P_\mu \rangle = \langle P_\mu, Q_\lambda \rangle.$

45

Combining (a), (b) and (c), we see the theorem is true.

4.4. Theorem 4.3 has some interesting consequences. For each element $F = \sum_{\nu \in X} f(\nu)e^{\nu} \in \Gamma$ and arbitrary $\lambda, \mu \in X^+$ we define

$$F_{\lambda,\mu} := \sum_{w \in W} \varepsilon(w)f(w(\lambda + \rho) - \mu - \rho).$$

Let θ be an element in X^+. Assume that $\theta + \nu = w(\pi + \rho) - \rho$ for some π in X^+, then $\nu = w(\pi + \rho) - \theta - \rho$ and $\chi_{\theta+\nu} = \varepsilon(w)\chi_{\pi}$. Therefore we have

(4.4.1)
$$\chi(e^{\theta}F) = \sum_{\pi \in X^+} F_{\pi,\theta}\chi_{\pi}.$$

In particular we have

(4.4.2)
$$P_{\lambda} = \eta_{\lambda}^{-1} \sum_{\pi \in X^+} \Phi_{\pi,\lambda}\chi_{\pi}.$$

(4.4.3)
$$Q_{\mu} = \sum_{\pi \in X^+} \Psi_{\pi,\mu}\chi_{\pi}.$$

Theorem 4.3 enables us to establish the relation between $\Psi_{\mu,\lambda}$ and $d_{\mu,\lambda}$ (defined in 3.15). We need some notations. Let $\mathcal{A}$ be the ring of Laurant polynomials in $\mathbf{q}_s^{\frac{1}{2}}$, $s \in S$. The map $\mathbf{q}_s^{\frac{1}{2}} \to \mathbf{q}_s^{-\frac{1}{2}}$ defines an involution $\bar{}$ of $\mathcal{A}$. Let $\alpha_1, ..., \alpha_k$ be simple roots of R and let $\beta = n_1\alpha_1 + \cdots + n_k\alpha_k$ be an element in $\mathbb{Z}R^+$, we set $\mathbf{q}^{\beta} = \mathbf{q}_{\alpha_1}^{n_1} \cdots \mathbf{q}_{\alpha_k}^{n_k}$. For $\lambda, \mu \in X$ we shall write $\mu \leq \lambda$ if $\lambda - \mu \in \mathbb{N}R^+$.

4.5. Theorem. Let λ, μ be arbitrary elements in X^+. Assume that $\mu \leq \lambda$, then

$$d_{\mu,\lambda} = \mathbf{q}^{\lambda-\mu}\overline{\Psi}_{\lambda,\mu}.$$

See 3.15 for the definition of $d_{\mu,\lambda}$. (Compare with [Ka2, Theorem 1.8].)

Proof. It is obvious that the equation system

$$\sum_{\substack{\pi \in X^+ \\ \mu \leq \pi \leq \lambda}} \eta_{\lambda}^{-1}\Phi'_{\lambda,\pi}A_{\pi,\mu} = \delta_{\lambda,\mu}, \qquad \lambda, \mu, \pi \in X^+ \text{ and } A_{\pi,\mu} \in \mathcal{A}$$

has a unique solution, where $\Phi'_{\lambda,\pi} = \Phi_{\pi,\lambda}$. It is easy to see that $\mathbf{q}^{\lambda-\mu}\bar{d}_{\mu,\lambda}$, $\mu, \lambda \in X^+$ and $\mu \leq \lambda$, is a solution (cf. 3.15). According to Theorem 4.3 we know that $\Psi_{\lambda,\mu}$ is also a solution of the equation system. Therefore $\mathbf{q}^{\lambda-\mu}\bar{d}_{\mu,\lambda} = \Psi_{\lambda,\mu}$. The theorem is proved.

Remark. According to the equality (4.3.1) and the definition of $\Psi_{\lambda,\mu}$ we may regard $\Psi_{\lambda,\mu}$ as a $\mathbf{q}_s$-analogue of Kostant's partition function.

46

4.6. Theorem. *For $\lambda \in X^+$ we have*

$$\chi_\lambda = \sum_{\mu \in X^+} \eta_\mu^{-1} \Psi_{\lambda,\mu} \sum_{w \in W} w(e^\mu \prod_{\alpha \in R^+} \frac{1 - \mathbf{q}_\alpha e^{-\alpha}}{1 - e^{-\alpha}}).$$

(Compare with [Ka2, Theorem 1.5].)

Proof. Using (4.4.2) and Theorem 4.3 we get $\chi_\lambda = \sum_{\mu \in X^+} \Psi_{\lambda,\mu} P_\mu$. From this equality and the definition of P_μ we see the theorem is true.

4.7. Denote S_1 the set of simple reflections in S correspomding short simple roots and S_2 the set of simple reflections in S correspomding long simple roots. (Convention: For type A_n, D_n, E_n all roots are regarded as short roots.) Let $\sigma_1 \in \mathbf{g}$ be the sum of root vectors corresponding to elements in S_1 and $\sigma_2 \in \mathbf{g}$ the sum of root vectors corresponding to elements in S_2. Let λ, μ be lements in X^+ and let $V(\lambda)$ be an irreducible $\mathbf{g}$-module of highest weight λ. Denote $V(\lambda)_\mu$ the μ-weight space of $V(\lambda)$. We may define a two-index filtration of $V(\lambda)_\mu$ as follows:

$$V(\lambda)_{\mu,i,j} := \{v \in V(\lambda)_\mu \mid \sigma_1^i v = 0 \text{ and } \sigma_2^j v = 0\}, \qquad i, j \in \mathbb{N}.$$

4.8. Question. *Keep the set up in 4.7. Does the equality*

$$d_{\mu,\lambda} = \sum_{i,j \in \mathbb{N}} (dim(V(\lambda)_{\mu,i,j}/V(\lambda)_{\mu,i-1,j}) + dim(V(\lambda)_{\mu,i,j}/V(\lambda)_{\mu,i,j-1}) \mathbf{u}^i \mathbf{v}^j$$

hold? where $\mathbf{u} = \mathbf{q}_s$ for any s in S_1 and $\mathbf{v} = \mathbf{q}_s$ for any s in S_2. (cf. [Br].)

5. Kazhdan-Lusztig Classification on Simple Modules of Affine Hecke Algebras

In this chapter we first recall some results of Ginsburg [G1-G2], Kazhdan and Lusztig [KL4], which will be needed later. We then give some discussions to the standard modules of affine Hecke algebras (in the sense of [KL4]). For type A_n it is not difficult to determine the dimensions of standard modules. In general it seems that one can determine the dimensions of standard modules by means of Green functions. A conjecture concerned with the based rings of two-sided cells in affine Weyl groups is stated. We also formulate a conjecture concerned with classifications of simple modules of affine Hecke algebras of two parameters, which is an analogue of the conjecture (*) in the introduction of the book.

Throughout this chapter all varieties and algebraic groups are over $\mathbb{C}$ except specified indications. We shall use algebraic (equivariant) K-theory instead of topological (equivariant) K-theory. See [T] for comparison between them.

For an algebraic group G and a G-variety M, we denote $K_G(M)$ the Grothendieck group of the category of G-equivariant coherent sheaves on M and $\mathbf{K}_G(M) = K_G(M) \otimes \mathbb{C}$ its complexification. The K-group $\mathbf{K}_G(M)$ has a natural $\mathbf{R}_G = \mathbf{K}_G(point)$ module structure. When $G = \{1\}$ is the unit group, we shall omit the subscript G of these K-groups.

We use $\mathcal{O}_M$ for the structure sheaf of M.

5.1. Convolution in K-theory. Following Ginsburg [G2] we consider the convolution in K-theory. In sections 5.1-5.3 we use the account in [GV].

Let G be an algebraic group over $\mathbb{C}$ and M_1, M_2, M_3 be smooth quasi-projective G-varieties. So the varieties $M_1 \times M_2 \times M_3$, $M_1 \times M_2$, $M_2 \times M_3$ and $M_1 \times M_3$ are also G-varieties and G acts on them diagonally. Then the natural projections $p_{ij} : M_1 \times M_2 \times M_3 \to M_i \times M_j$ commute with G-actions, i.e., they are G-equivariant morphisms. Let Z be a G-stable closed subvariety of $M_1 \times M_2$ and $\tilde{Z}$ be a G-stable closed subvariety of $M_2 \times M_3$. Assume that the morphism

$$(5.1.1) \qquad p_{13} : p_{12}^{-1}(Z) \cap p_{23}^{-1}(\tilde{Z}) \to M_1 \times M_3 \quad \text{is proper.}$$

Then its image is a G-stable closed subvariety of $M_1 \times M_3$ and will be called the composition of Z and $\tilde{Z}$. We denote it by $Z \circ \tilde{Z}$. Following Ginsburg we define the convolution map (see [G2])

$$(5.1.2) \qquad * : K_G(Z) \otimes K_G\tilde{Z} \to K_G(Z \circ \tilde{Z})$$

as follows. Let $[\mathcal{F}] \in K_G(Z)$ and $[\tilde{\mathcal{F}}] \in K_G(\tilde{Z})$ be the classes of some G-equivariant coherent sheaves $\mathcal{F}$ on Z and $\tilde{\mathcal{F}}$ on $\tilde{Z}$. Set

$$(5.1.3) \qquad \mathcal{F} * \tilde{\mathcal{F}} = (Rp_{13})_* (p_{12}^* \mathcal{F} \overset{L}{\underset{\mathcal{O}_{M_1 \times M_2 \times M_3}}{\otimes}} p_{23}^* \tilde{\mathcal{F}}).$$

48

In this formula the upper star stands for the pullback morphism, well-defined for smooth maps (see, e.g. [Fu]), for example, $p_{12}^* \mathcal{F} = \mathcal{F} \boxtimes \mathcal{O}_{M_3}$. To define $\overset{L}{\otimes}$, on $M_1 \times M_2 \times M_3$ we choose a finite resolution $\mathbf{F}_{12}^{\cdot}$ of $p_{12}^* \mathcal{F}$ and a finite resolution $\mathbf{F}_{23}^{\cdot}$ of $p_{23}^* \tilde{\mathcal{F}}$ by G-equivariant locally free sheaves, which exist since $M_1 \times M_2 \times M_3$ is smooth and quasi-projective (see [T, 1.9 (c)]). The simple complex $\mathbf{F}_{12} \otimes \mathbf{F}_{23}$ associated to the tensor product of the resolutions represents the tensor product $\overset{L}{\otimes}$ in a derived category. This complex is exact off $p_{12}^{-1}(Z) \cap p_{23}^{-1}(\tilde{Z})$. Hence, its derived direct image $\mathcal{F} * \tilde{\mathcal{F}} = (Rp_{13})_* (\mathbf{F}_{12} \otimes \mathbf{F}_{23})$ is a complex of sheaves on $M_1 \times M_3$ whose cohomology sheaves are coherent sheaves on $M_1 \times M_3$ since the assumption (5.1.1); moreover, these cohomology sheaves are supported on $Z \circ \tilde{Z}$. We let $[\mathcal{F}] * [\tilde{\mathcal{F}}] \in K_G(Z \circ \tilde{Z})$ be the alternating sum of these cohomology sheaves. The definition of $*$ doesnot depend on the choices involved. Furthermore, the convolution is associative in a natural way.

5.2. From now on we assume that G is a reductive group and s a semisimple element in G. Evaluating a character of the group G at the element s gives rise to an algebra homomorphism: $\mathbf{R}_G \to \mathbb{C}$. Let $\mathbb{C}_s$ denote the one-dimensional $\mathbf{R}_G$-module arising from the homomorphism. It is known that each simple $\mathbf{R}_G$-module is isomorphic to some $\mathbb{C}_s$, and $\mathbb{C}_s$ is isomorphic to $\mathbb{C}_t$ if and only if s, t are conjugate in G.

Let M be a smooth G-variety and let M^s be the s-fixed points subvariety. By the slice theorem [Lu], M^s is smooth. Let $\mathcal{N}_s^*$ denote the conormal sheaf at the subvariety $M^s \hookrightarrow M$. The s-action on M induces a natural s-action on $\mathcal{N}_s^*$. We set

$$\lambda(M^s) = \sum (-1)^k \mathrm{tr}(s, \overset{k}{\wedge} \mathcal{N}_s^*) \in \mathbf{K}(M^s).$$

Here $\mathbf{K}(M^s)$ is the ordinary K-group; and for a vector bundle E with semisimple s-action on the fibres, we use the notation $\mathrm{tr}(s, E) = \sum a_k \cdot E_k$, where a_k are the eigenvalues of the action s and E_k stands for the subbundle of E corresponding to the eigenvalue a_k. The element $\lambda(M^s)$ is invertible in $\mathbf{K}(M^s)$ since all the eigenvalues of the action s are not equal to zero.

The inclusion of varieties $i : M^s \hookrightarrow M$ gives rise to the direct image functor $i_! : \mathbf{K}(M^s) \to \mathbf{K}_G(M)$ and to the inverse image functor $i^! : \mathbf{K}_G(M) \to \mathbf{K}(M^s)$. The later is defined by the formula

$$i^! \mathcal{F} = \sum (-1)^k \mathrm{Tor}_{\mathcal{O}_M}^k (\mathcal{F}, \lambda(M^s)^{-1}).$$

The morphism $i^!$ clearly factors through the quotient $\mathbb{C}_s \otimes_{\mathbf{R}_G} \mathbf{K}_G(M)$. One has the following localization theorem (see [T])

(a) (i). $i^! \circ i_! = \mathrm{Id}_{\mathbf{K}(M^s)}$. In particular, the morphism below is surjective:

$$i^! : \ \mathbb{C}_s \otimes_{\mathbf{R}_G} \mathbf{K}_G(M) \twoheadrightarrow \mathbf{K}(M^s)$$

(ii). This morphism is an isomorphism provided the group G is abelian.

49

Keep the notations G and s. Let M_1, M_2, M_3 be smooth G-varieties and assume that Z is a closed G-subvariety of $M_1 \times M_2$ and let $j : Z \hookrightarrow M_1 \times M_2$ be the inclusion. Define a morphism $\mathbf{r}_s : \mathbf{K}_G(Z) \to \mathbf{K}(Z^s)$ by the formula

$$(5.2.1) \qquad \mathcal{F} \longmapsto \mathbf{r}_s(\mathcal{F}) = \sum (-1)^k \mathrm{Tor}^k_{\mathcal{O}_{M_1 \times M_2}}(\lambda(M_1^s)^{-1} \boxtimes \mathcal{O}_{M_2^s}, j_! \mathcal{F}).$$

The Tor groups on the right hand side are supported on Z^s, since $Z^s = Z \cap (M_1 \times M_2)$. The assignment $\mathcal{F} \longmapsto \mathbf{r}_s(\mathcal{F})$ factors through $\mathbb{C}_s \otimes_{\mathbf{R}_G} \mathbf{K}_G(Z)$. So we get a morphism $\mathbf{r}_s : \mathbb{C}_s \otimes_{\mathbf{R}_G} \mathbf{K}_G(Z) \to \mathbf{K}(Z^s)$. Let $\tilde{Z}$ be a closed G-subvariety of $M_2 \times M_3$. Similarly we have a morphism $\mathbf{r}_s : \mathbb{C}_s \otimes_{\mathbf{R}_G} \mathbf{K}_G(\tilde{Z}) \to \mathbf{K}(\tilde{Z}^s)$.

(b) Bivariant fixed-point theorem. Assume that $Z, \tilde{Z}$ satisfy (5.1.1), then the following diagram

$$
\begin{array}{ccc}
(\mathbb{C}_s \otimes_{\mathbf{R}_G} \mathbf{K}_G(Z)) \otimes (\mathbb{C}_s \otimes_{\mathbf{R}_G} \mathbf{K}_G(\tilde{Z})) & \longrightarrow & \mathbb{C}_s \otimes_{\mathbf{R}_G} \mathbf{K}_G(Z \circ \tilde{Z}) \\
{\scriptstyle \mathbf{r}_s} \downarrow & & \downarrow {\scriptstyle \mathbf{r}_s} \\
\mathbf{K}(Z^s) \otimes \mathbf{K}(\tilde{Z}^s) & \longrightarrow & \mathbf{K}(Z^s \circ \tilde{Z}^s)
\end{array}
$$

commutes. That is, the convolution commutes with the the morphism $\mathbf{r}_s$.

The K-theoretic convolution has its counterpart in homology. For $Z \subset M_1 \times M_2$, the complex coeffcients Borel-Moore homology group $H_i(Z)$ may be defined (via Poincaré duality) as the relative cohomology $H^{m-i}(M_1 \times M_2, M_1 \times M_2 \setminus Z)$, where $m = \dim(M_1 \times M_2)$. The cup-product in cohomology gives rise to a cap-product on Borel-Moore homology, which replaces the functor $\overset{L}{\otimes}$ in K-theory. One defines a homology counterpart of (5.1.2) as a map

$$* : H_i(Z) \otimes H_j(\tilde{Z}) \to H_{i+j-d}(Z \circ \tilde{Z}), \quad d = \dim M_2,$$

given by the formula: $c * \tilde{c} = (p_{13})_*(p_{12}^* c \cap p_{23}^* \tilde{c})$.

Associate an element $\mathcal{F} \in \mathbf{K}(Z)$ to its *Chern character* class $ch(\mathcal{F}) \in H_*(Z)$ (see [FM]). Further, let $\mathrm{Td}(M_2) \in H_*(M_2)$ denote the *Todd class* of the manifold M_2. Define a morphism $\mathbf{c} : \mathbf{K}(Z) \to H_*(Z)$ by the formula

$$\mathbf{c}(\mathcal{F}) = pr_2^* \mathrm{Td}(M_2) \cdot ch(\mathcal{F}),$$

where pr_2 is the second projection $M_1 \times M_2 \to M_2$.

(c) Bivariant Riemann-Roch theorem. The morphism $\mathbf{c}$ commutes with the convolution. That is, the diagram

$$
\begin{array}{ccc}
\mathbf{K}(Z) \otimes \mathbf{K}(\tilde{Z}) & \longrightarrow & \mathbf{K}(Z \circ \tilde{Z}) \\
{\scriptstyle \mathbf{c}} \downarrow & & \downarrow {\scriptstyle \mathbf{c}} \\
H_*(Z) \otimes H_*(\tilde{Z}) & \longrightarrow & H_*(Z \circ \tilde{Z})
\end{array}
$$

commutes.

Combining (c) and (d) (applied to Z^s), we obtain the following result.

(d) The composition morphism $\mathbf{c} \circ \mathbf{r}_s : \mathbb{C}_s \otimes_{\mathbf{R}_G} \mathbf{K}_G(Z) \to H_*(Z^s)$ commutes with convolution.

5.3. Convolution algebra.   Let G be an algebraic group, M a smooth quasi-projective G-variety. Let $\pi : M \to N$ be a G-equivariant proper morphism. Set

$$Z = M \times_N M = \{(m, m') \in M \times M \mid \pi(m) = \pi(m')\} \subset M \times M.$$

We view Z as a G-equivariant correspondence in $M \times M$. Clearly we have $Z \circ Z = Z$. Thus the convolution makes $\mathbf{K}_G(Z)$ an associative $\mathbf{R}_G$-algebra. Observe further that the diagonal of $M \times M$ is contained in Z and the class of the structure sheaf of the diagonal is the unit of the algebra $\mathbf{K}_G(Z)$. Similarly, the convolution makes the Borel-Moore homology group $H_*(Z)$ into a finite dimensional $\mathbb{C}$-algebra whose unit is the fundamental class of the diagonal.

5.4. Examples of convolution algebra.   Assume that M is a projective variety and $N = \{point\}$. Let $\pi : M \to N$ be the unique morphism from M to N. Then we have $Z = M \times M$. We assume that G is reductive (possibly disconnected). For any semisimple element $s \in G$, by 5.2(d) we have an algebra homomorphism

$$(5.4.1) \qquad \mathbf{c} \circ \mathbf{r}_s : \mathbb{C}_s \otimes_{\mathbf{R}_G} \mathbf{K}_G(M \times M) \to H_*(M^s \times M^s)$$

Now the algebra $H_*(M^s \times M^s)$ and the group $A(s) = C_G(s)/C_G(s)^0$ naturally act on the homology group $H_*(M^s)$, and the actions commute. Let ρ be a simple $A(s)$-module which appears in the homology group $H_*(M^s)$. Let

$$E_{s,\rho} = (\rho^* \otimes H_*(M^s))^{A(s)} = \mathrm{Hom}_{A(s)}(\rho, H_*(M^s)),$$

where ρ^* is the dual of ρ. The space $E_{s,\rho}$ is in fact a simple $H_*(M^s \times M^s)$-module and by varying ρ one gets each simple $H_*(M^s \times M^s)$-module exactly once. Furthermore, we can regard $E_{s,\rho}$ as a $\mathbf{K}_G(M \times M)$-module, via $\mathbf{c} \circ \mathbf{r}_s$; this is a simple $\mathbf{K}_G(M \times M)$-module and $(s, \rho) \to E_{s,\rho}$ defines a bijection between the set of pairs (s, ρ) as above (up to G-conjugacy) and the set of isomorphism classes of simple $\mathbf{K}_G(M \times M)$-modules.

Let $i : M \times M \to M \times M$ be the G-equivariant morphism defined by $i(m, m') = (m', m)$. For any G-equivariant coherent sheaf $\mathcal{F}$, we shall write $\tilde{\mathcal{F}}$ for $i^!(\mathcal{F}^*)$, here $\mathcal{F}^*$ is the dual sheaf of $\mathcal{F}$. We write E_s for $E_{s,\rho}$ when ρ is the unit representation of $C_G(s)/C_G(s)^0$.

When both G and M are finite, the algebra $\mathbf{K}_G(Z)$ is just that defined in [L15]. When G is reductive and M is finite, the algebra $\mathbf{K}_G(Z)$ is just that defined in [L14]. When G acts trivially on a finite set M, then the convolution algebra $\mathbf{K}_G(M \times M)$

is isomorphic to the algebra $M_{k \times k}(\mathbf{R}_G)$, the $k \times k$ matrix ring over $\mathbf{R}_G$, where $k = |M|$.

Another example is that an affine Hecke algebra can be realized as a convolution algebra. This is a key of Kazhdan & Lusztig's work on classification of simple modules of affine Hecke algebras (see [KL4]). We shall recall their work in next section.

Assume that Y is a finite G-variety, i.e., a finite G-set. Any coherent sheaf on Y is a vector bundle (v.b.) on Y. If V is an irreducible G-v.b. on Y, then the set $\{y \in Y \mid V_y \neq 0\}$ is a single G-orbit $\mathcal{O}$ in Y, and for any $y \in \mathcal{O}$, the obvious representation of the isotropy group G_y on V_y is irreducible; this gives a bijection between the set of irreducible G-v.b. on Y (up to isomorphisms) and the set of pairs (y, ρ) where $y \in Y$, ρ is an irreducible (algebraic) representation of G_y, modulo the obvious action of G. In chapters 11 and 12 we often write ρ_y for the irreducible G-v.b. on Y corresponding to the pair (y, ρ).

5.5. Geometric realization of affine Hecke algebra. In this section we shall assume that G is a connected algebraic group with simply connected derived group. Let $\mathbf{g}$ be the Lie algebra of G and let $\mathcal{N}$ be the variety of all nilpotent elements of $\mathbf{g}$. Denote $\mathcal{B}$ the variety of all Borel subalgebras of $\mathbf{g}$. Set

$$\tilde{\mathcal{N}} := \{(N, \mathbf{b}) \in \mathcal{N} \times \mathcal{B} \mid N \in \mathbf{b}\}$$

and let $\mu : \tilde{\mathcal{N}} \to \mathcal{N}$ be the Springer resolution by projecting $(N, \mathbf{b})$ to N. Let

$$Z = \tilde{\mathcal{N}} \times_{\mathcal{N}} \tilde{\mathcal{N}} \simeq \{(N, \mathbf{b}, \mathbf{b}') \mid N \in \mathbf{b} \cap \mathbf{b}' \text{ is nilpotent, } \mathbf{b}, \mathbf{b}' \in \mathcal{B}\}$$

be the Steinberg variety. The Steinberg variety can be obtained in another way. The group G acts on $\mathbf{g}$ through adjoint action, so G acts on the variety $\mathcal{B}$. Let G act on $\mathcal{B} \times \mathcal{B}$ diagonally, then the number of G-orbits in $\mathcal{B} \times \mathcal{B}$ is $|W_0|$. Let $T^*(\mathcal{B} \times \mathcal{B})$ be the cotangent bundle of $\mathcal{B} \times \mathcal{B}$, then the union of conormal bundles of all G-orbits in $\mathcal{B} \times \mathcal{B}$ is isomorphic to Z.

Convention: For any $g \in G$, $x \in \mathbf{g}$ and $\mathbf{b} \in \mathcal{B}$, we shall write $g.x$ and $g.\mathbf{b}$ instead of $\mathrm{Ad}g(x)$ and $\mathrm{Ad}g(\mathbf{b})$ respectively.

Let $G \times \mathbb{C}^*$ act on $\tilde{\mathcal{N}}$ by

$$(5.5.1) \qquad\qquad (g, q) : (N, \mathbf{b}) \longmapsto (g.q^{-1}N, g.\mathbf{b}).$$

Then $G \times \mathbb{C}^*$ acts on Z by

$$(5.5.2) \qquad\qquad (g, q) : (N, \mathbf{b}, \mathbf{b}') \longmapsto (g.q^{-1}N, g.\mathbf{b}, g.\mathbf{b}').$$

Keep the notations in 2.1-2.2. Let $\mathbf{H} = H \otimes_A \mathbb{C}[\mathbf{q}^{\frac{1}{2}}, \mathbf{q}^{-\frac{1}{2}}]$. We shall identify $\mathbf{A} = \mathbb{C}[\mathbf{q}, \mathbf{q}^{-1}]$ with $\mathbf{R}_{\mathbb{C}^*}$ by regarding $\mathbf{q}$ as the identity representation $\mathbb{C}^* \to \mathbb{C}^*$. Let $\dot{\mathbf{H}}$ be the $\mathbf{A}$-subalgebra of $\mathbf{H}$ generated by T_w, θ_x, $w \in W$, $x \in X$. We shall identify the center of $\dot{\mathbf{H}}$ with the ring $\mathbf{R}_{G \times \mathbb{C}^*}$ by regarding U_x, $x \in X^+$ as the the irreducible module of highest weight x. Then (see [KL4], see also [G2])

52

(a) There exists an $\mathbf{R}_{G\times\mathbb{C}^*}$-algebra isomorphism between the affine Hecke algebra $\dot{\mathbf{H}}$ and the convolution algebra $\mathbf{K}_{G\times\mathbb{C}^*}(Z)$.

5.6. Standard modules. Let G be as in 5.5. Given a semisimple element $(s, q) \in G \times \mathbb{C}^*$, set

$$\mathbf{g}_{s,q} = \{\xi \in \mathbf{g} \mid s.\xi = q\xi\},$$
$$\mathcal{N}_{s,q} = \{\xi \in \mathcal{N} \mid s.\xi = q\xi\}.$$

For each $N \in \mathcal{N}_{s,q}$, consider the variety

$$\mathcal{B}^s_N = \{\mathbf{b} \in \mathcal{B} \mid N \in \mathbf{b} \text{ and } s.\mathbf{b} = \mathbf{b}\}.$$

Obviously the group $C_G(s) \cap C_G(N)$ acts on the variety $\mathcal{B}^s_N$. So the group $A(s, N) = C_G(s) \cap C_G(N)/(C_G(s) \cap C_G(N))^0$ acts on the homology group $H_*(\mathcal{B}^s_N)$. Let $A(s, N)^\vee$ be the set of isomorphism classes of the irreducible $A(s, N)$-modules which appear in $H_*(\mathcal{B}^s_N)$. It is shown in [KL4] that

(a) There exists an $\dot{\mathbf{H}}$-module structure on $\mathbf{K}(\mathcal{B}^s_N) = H_*(\mathcal{B}^s_N)$ such that

   (i). The action of $\dot{\mathbf{H}}$ commutes with the action of $A(s, N)$.

   (ii). $\mathbf{q}$ acts on it by scalar q and $U_x, x \in X^+$ acts on it by scalar $tr(s, V(x))$, where $V(x)$ is the irreducible G-module with highest weight x.

For any $\rho \in A(s, N)^\vee$, define

$$M_{s,N,q,\rho} := \mathrm{Hom}_{A(s,N)}(\rho, \mathbf{K}(\mathcal{B}^s_N)) = (\mathbf{K}(\mathcal{B}^s_N) \otimes \rho^*)^{A(s,N)},$$

where ρ^* is the dual module of ρ. By (a) we see that $M_{s,N,q,\rho}$ is an $\dot{\mathbf{H}}$-module. The $\dot{\mathbf{H}}$-module $M_{s,N,q,\rho}$ is called a standard module. We say that two quadruples (s, N, q, ρ) and (s', N', q', ρ') are G-conjugate if there exists some $g \in G$ such that $s' = gsg^{-1}$, $N' = g.N$, $q = q'$, $\rho' = g(\rho)$ (note that we have a natural bijection $g : A(s, N)^\vee \to A(s', N')^\vee$ when the first two conditions hold). Now we can state the main results of [KL4].

(b) Let L be a simple module of $\dot{\mathbf{H}}$ such that $\mathbf{q}$ acts on it by scalar q, then L is a quotient module of some standard module $M_{s,N,q,\rho}$.

(c) Two standard modules $M_{s,N,q,\rho}$ and $M_{s',N',q',\rho'}$ are isomorphic if and only if (s, N, q, ρ) and (s', N', q', ρ') are G-conjugate.

(d) When q is not a root of 1, then each standard module $M_{s,N,q,\rho}$ has a unique quotient module, denote by $L_{s,N,q,\rho}$. Moreover, $L_{s,N,q,\rho}$ and $L_{s',N',q,\rho'}$ are isomorphic if and only if (s, N, q, ρ) and (s', N', q, ρ') are G-conjugate.

Lusztig conjecture that (d) remains true provided that $\sum_{w \in W_0} q^{l(w)} \neq 0$. (See [L17].)

Given $q \in \mathbb{C}^*$, regard $\mathbb{C}$ as an $\mathbf{A}$-algebra (resp. $\mathbb{C}[\mathbf{q}^{\frac{1}{2}}, \mathbf{q}^{-\frac{1}{2}}]$-algebra) by specializing $\mathbf{q}$ to q (resp. $\mathbf{q}^{\frac{1}{2}}$ to a square root of q), then consider the Hecke algebra

$$\mathbf{H}_q = \dot{\mathbf{H}} \otimes_{\mathbf{A}} \mathbb{C} = \mathbf{H} \otimes_{\mathbb{C}[\mathbf{q}^{\frac{1}{2}}, \mathbf{q}^{-\frac{1}{2}}]} \mathbb{C}.$$

For each semisimple element s in G, let $\mathbf{I}_{s,q}$ be the two-sided ideal of $\mathbf{H}_q$ generated by $U_x - tr(s, V(x))$, $x \in X^+$ and let $\mathbf{H}_{s,q}$ be the quotient algebra $\mathbf{H}_q/\mathbf{I}_{s,q}$ of $\mathbf{H}_q$. It is easy to see that $\dot{\mathbf{H}}$ acts on the standard module $M_{s,N,q,\rho}$ factoring through the algebra $\mathbf{H}_{s,q}$. The following result is due to Ginzburg, which can be deduced from 5.2(d) and 5.5(a).

(e) The algebra $\mathbf{H}_{s,q} = \mathbb{C}_{s,q} \otimes \mathbf{K}_{G \times \mathbb{C}^*}(Z)$ is isomorphic to the convolution algebra $H_*(Z^{s,q})$.

  (By definition, $Z^{s,q} = \{(N, \mathbf{b}, \mathbf{b}') \in Z \mid s.N = qN,\ s.\mathbf{b} = \mathbf{b},\ \text{and}\ s.\mathbf{b}' = \mathbf{b}'\}$.)

The above results yield naturally two questions.

5.7. Question. (i). *Determine the dimensions of the standard modules $M_{s,N,q,\rho}$ and of the simple modules $L_{s,N,q,\rho}$ when q is not a root of 1.*

(ii). *Classify the simple modules of $\mathbf{H}_q$ when q is a root of 1.*

For the question 5.7 (ii) in next chapter we will show that 5.6(d) is true for most roots of 1 by Combining 5.6(d-e). It is difficult to get the dimensions of simple $\mathbf{H}_q$-modules. But we can say a few words for the dimensions of standard modules.

5.8. Dimensions of certain standard modules. The following results can be used to calculate the dimensions of certain standard modules.

(a) $H_{odd}(\mathcal{B}_N^s) = 0$ and $H_{even}(\mathcal{B}_N^s)$ is isomorphic to the Chow group of $\mathcal{B}_N^s$ (with complex coefficients). (See [CLP]).

(b) Assume that a connected diagonalizable algebraic group D acts on a variety M, then $\chi(M) = \chi(M^D)$, where $\chi(\cdot)$ denotes the Euler number (see [BB], I am grateful to Professor R.V. Gurjar for providing the reference).

Notations are as in 5.5. For any semisimple element $(s, q) \in G \times \mathbb{C}^*$, we always have $s.0 = 0 = q0$. Since G has a simply connected derived group, so $C_G(s)$ is connected. Thus $A(s, 0)^\vee$ only contains the unit representation, denote by 1. Moreover we have $M_{s,0,q,1} = \mathbf{K}(\mathcal{B}^s) = H_*(\mathcal{B}^s)$. By (a) we know that

$$(5.8.1) \qquad \dim M_{s,0,q,1} = \chi(\mathcal{B}^s).$$

Let T_s be a maximal torus of G containing s. Then T_s acts on $\mathcal{B}^s$. Using (b) we see that $\chi(\mathcal{B}^s) = \chi(\mathcal{B}^{T_s})$. It is well known that $\mathcal{B}^{T_s}$ is a finite set of $|W_0|$ elements. Thus we get

$$(5.8.2) \qquad \dim M_{s,0,q,1} = |W_0|.$$

One also can obtain (5.8.2) by using 5.9(c) and results in [X2].

Now we assume that $G = SL_n(\mathbb{C})$. The results (a-b) are also sufficient to determine the dimensions of standard modules of $\mathbf{H}_q$ in this case. Let $(s, q) \in G \times \mathbb{C}^*$ be a semisimple element, it is harmless to assume that $(s, q) \in T \times \mathbb{C}^*$, where T is the

54

subgroup of G consisting of diagonal matrices in G. Let $N \in \mathcal{N}_{s,q}$. Noting that $\mathbf{g} = sl_n(\mathbb{C})$, we see that the sizes of the Jordan blocks of N determine a partition of n, denoted λ. It is known that $A(s, N)^\vee$ only contains the unit representation, also denote by 1. Moreover we have $M_{s,N,q,1} = \mathbf{K}(\mathcal{B}_N^s) = H_*(\mathcal{B}_N^s)$. By (a) we know that

$$(5.8.3) \qquad \dim M_{s,N,q,1} = \chi(\mathcal{B}_N^s).$$

A direct calculation shows that the pair (s, N) is conjugate to certain pair (s', N') such that $s' \in T$ and $N' = \text{diag}(N_1', N_2', ..., N_k')$, where each N_i' is a Jordan block. We may choose all N_i' to be upper triangular matrices. It is no harm to assume that $(s, N) = (s', N')$. It is known (also easy to check) that $T_N = (T \cap C_G(N))^0$ is a maximal torus in the group $C_G(N)$. Since $T_N \subset C_G(s) \cap C_G(N)$, T_N acts on the variety $\mathcal{B}_N^s$. Using (b) we know that $\chi(\mathcal{B}_N^s) = \chi((\mathcal{B}_N^s)^{T_N})$. Let $t \in T_N$ be a regular element in $C_G(N)$, then $g = t \exp N$ is a regular element in G. According to [St1] the variety $(\mathcal{B}_N^s)^{T_N} = B^g$ is finite and

(5.8.4) *The cardinal $|(\mathcal{B}_N^s)^{T_N}|$ of $(\mathcal{B}_N^s)^{T_N}$ is the number of elements $w \in W_0$ such that $w.N \in \mathbf{b}$, where $\mathbf{b}$ is the Borel subalgebra of $\mathbf{g}$ consisting of all upper triangular matrices in $\mathbf{g}$.*

Let $\mu = (\mu_1, \mu_2, ..., \mu_k)$ be the dual partition of λ, using (5.8.4) we get

$$(5.8.5) \qquad \dim M_{s,N,q,1} = \frac{n!}{\mu_1! \mu_2! \cdots \mu_k!}.$$

Note that $\dim M_{s,N,q,1}$ is the number of left cells containing in the two-sided cell of W corresponding to the nilpotent G-orbit containing N.

For other types we can get similar results when $C_G(N)$ is connected.

Finally we state a result concerned with the relations between $M_{s,N,q,\rho}$ and $M_{s,0,q,1}$.

(c). *The injection $\mathcal{B}_N^s \hookrightarrow \mathcal{B}_N$ induces an $\mathbf{H}_q$-module injection $M_{s,N,q,\rho} \hookrightarrow M_{s,0,q,1}$.*

For type A this was proved in [HS]. In general it follows from the results in [KL4, CLP].

For type A, we can get a little more. Let $s, t \in T$, and $N \in \mathcal{N}_{s,q}$, $N' \in \mathcal{N}_{t,q}$. Assume that $\mathcal{B}_{N'}^t \subseteq \mathcal{B}_N^s$, then the inclusion induces an $\mathbf{H}_q$-module injection: $M_{t,N',q,1} \hookrightarrow M_{s,N,q,1}$. The proof is similar to that in [HS]. This fact should be useful for calculating the multiplicities of simple modules in standard modules. The multiplicities should have a nice combinatorial description.

5.9. The asymptotic Hecke algebra J. The work [L12-L14] shows that the asymptotic Hecke algebra $\mathbf{J} = J_{\mathbb{Z}} \otimes \mathbb{C}$ of W is interesting in representation theory of affine Hecke algebras. Let $\phi_q : \mathbf{H}_q \to \mathbf{J}$ be the homomorphism induced from the homomorphism ϕ in 2.7(c). Thus each $\mathbf{J}$-module E gives rise to an $\mathbf{H}_q$-mdoule E_q via ϕ_q. We shall need the following result.

(a) There is a unique $\mathbb{C}$-algebra involution $\psi : \mathbf{H}_q \to \mathbf{H}_q$ such that $\psi(T_s) = -qT_s^{-1}$, $\psi(\theta_x) = \theta_x^{\pm 1}$, $s \in S \cap W_0$, $x \in X$. (See [KL4].)

The involution k in 1.6 (e) induces an involution of the $\mathbb{C}$-algebra $\mathbf{H}_q$, denoted again k. For each $\mathbf{H}_q$-module M we may define a new $\mathbf{H}_q$-module structure on M through the automorphism $k\psi$ of $\mathbf{H}_q$, denoted M^* the new $\mathbf{H}_q$-module structure on M.

From 2.7(d) we know that (see [L13])

(b) The algebra $\mathbf{J}$ is finitely generated over its center. In particular, each simple $\mathbf{J}$-module is of finite dimension over $\mathbb{C}$.

We recall that there is a bijection between Cell(W) and the set of nilpotent G-orbits in $\mathbf{g}$. For $c \in$ Cell(W), we denote $\mathcal{N}_c$ the corresponding nilpotent G-orbit.

For each two-sided cell c of W, define $\mathbf{J}_c := J_{\mathbb{Z},c} \otimes \mathbb{C}$. Then $\mathbf{J} = \bigoplus_{c \in \mathrm{Cell}(W)} \mathbf{J}_c$. For every simple module E of $\mathbf{J}$, there exists a unique two-sided cell c of W such that $\mathbf{J}_c E \neq 0$. We call c the two-sided cell attached to E and denote it by c_E. We also call E a simple $\mathbf{J}$-module attached to c. Similarly, for each simple module L of $\mathbf{H}_q$, there exists a unique two-sided cell c of W such that

$$C_w L = 0 \quad \text{when } w \notin c \text{ and } w \underset{LR}{\leq} u \text{ for some } u \in c,$$

$$C_u L \neq 0 \quad \text{for some } u \in c,$$

we call c the two-sided cell attached to L and denote it by c_L. We also call L a simple $\mathbf{H}_q$-module attached to c.

(c) Let E be a simple $\mathbf{J}$-module. Then E_q^* is isomorphic to certain standard module $M_{s,N,q,\rho}$, $N \in \mathcal{N}_{c_E}$. Each standard $\mathbf{H}_q$-module can be obtain in this way. (See [L14]).

(d) Let E be as in (c). For each simple constituent L' of E_q, we have $c_E \underset{LR}{\leq} c_{L'}$.

(e) For each simple $\mathbf{H}_q$-module L, there exists some simple $\mathbf{J}$-module E such that

(i). $c_E = c_L$.

(ii). L is a simple quotient of E_q.

(iii). For any other simple constituent L' of E_q, we have $c_L \underset{LR}{\leq} c_{L'}$, $c_L \neq c_{L'}$.

(f) Assume that q is not a root of 1 or $q=1$. Then for each simple $\mathbf{J}$-module E, the $\mathbf{H}_q$-module E_q has a unique simple constituent L such that $c_L = c_E$, which is the unique quotient of E_q, denoted L_E. The map $E \to L_E$ defines a bijection between the set of isomorphism classes of simple $\mathbf{J}$-modules and the set of isomorphism classes of simple $\mathbf{H}_q$-modules.

We shall denote $Y_{q,c}$ the set of isomorphism classes of simple $\mathbf{H}_q$-modules to which the attached two-sided cell is c. For a semisimple element s in G, we denote $Y_{s,q}$

the set of isomorphism classes of simple $\mathbf{H}_q$-modules on which U_x ($x \in X^+$) acts by scalar $tr(s, V(x))$. Set $Y_{s,q,c} = Y_{q,c} \cap Y_{s,q}$.

The results (c-e) have several consequences. We need a simple fact. For each semisimple element $(s, q) \in G \times \mathbb{C}^*$, there exists a unique nilpotent G-orbit $\mathbf{n}_{s,q}$ such that $\bar{\mathbf{n}}_{s,q} \supseteq \mathcal{N}_{s,q}$ and for other nilpotent orbit $\mathbf{n}$ if $\bar{\mathbf{n}} \supseteq \mathcal{N}_{s,q}$, then $\bar{\mathbf{n}} \supseteq \mathbf{n}_{s,q}$.

5.10. Corollary. *For arbitrary* $N \in \mathbf{n}_{s,q}$ *and* $\rho \in A(s, N)^\vee$, *the module* $M_{s,N,q,\rho}$ *is simple.*

We conjecture that $M_{s,N,q,1}$ is simple if and only if $N \in \mathbf{n}_{s,q}$. When $\rho \neq 1$, the module $M_{s,N,q,\rho}$ may be simple for $N \notin \mathbf{n}_{s,q}$, see the dimension of E_4 in the part (G) of 12.3.

5.11. Corollary. *Assume that* $\mathcal{N}_{s,q} = \{0\}$, *then* $M_{s,0,q,1}$ *is simple and* $Y_{s,q} = \{M_{s,0,q,1}\}$.

Note that $\dim M_{s,0,q,1} = |W_0|$.

For each root α in R we denote $\mathbf{g}_\alpha$ the corresponding root subspace of $\mathbf{g}$.

5.12. Corollary. *Assume that* $\mathbf{g}_{s,q} \subseteq \mathbf{g}_\alpha \oplus \mathbf{g}_{-\alpha}$ *for some* $\alpha \in R$ *and* $A(s, N)^\vee = \{1\}$ *for some nonzero element* N *in* $\mathcal{N}_{s,q}$. *(Note that* N, N' *are conjugate under* $C_G(s)$ *for any* $N, N' \in \mathcal{N}_{s,q}$ *in our case). Then* $|Y_{s,q}| \leq 2$, *and*

(5.12.1). $\dim L_{s,N,q,1} = |W_0|/2$, *where* $0 \neq N \in \mathcal{N}_{s,q}$.

(5.12.2). *If* $|Y_{s,q}| = 2$, *we have* $\dim L_{s,0,q,1} = |W_0|/2$.

5.13. Let $K(\mathbf{H}_q)$ (resp. $K(\mathbf{J})$) be the Grothendieck group of $\mathbf{H}_q$-mdoules (resp. $\mathbf{J}$-modules) of finite dimensions over $\mathbb{C}$. According to 5.9(e), the map $E \to E_q$ defines a surjection $(\phi_q)_*: K(\mathbf{J}) \to K(\mathbf{H}_q)$ (see [L13]).

For a two-sided cell c of W, we denote $K(\mathbf{H}_q)_c$ the subgroup of $K(\mathbf{H}_q)$ spanned by those simple $\mathbf{H}_q$-modules L with $c_L = c$. Then we have $K(\mathbf{H}_q) = \oplus_c K(\mathbf{H}_q)_c$, where c runs over the set of two-sided cells in W. It is obvious that $K(\mathbf{J}) = \oplus_c K(\mathbf{J}_c)$, where the definition of $K(\mathbf{J}_c)$ is similar to that of $K(\mathbf{J})$. By the definition of ϕ_q we see that $(\phi_q)_*$ is compatible with the filtrations

$$K(\mathbf{J})_{\geq c} = \bigoplus_{c \underset{LR}{\leq} c'} K(\mathbf{J}_{c'}), \qquad K(\mathbf{H}_q)_{\geq c} = \bigoplus_{c \underset{LR}{\leq} c'} K(\mathbf{H}_q)_{c'}$$

of $K(\mathbf{J})$, $K(\mathbf{H}_q)$, hence $(\phi_q)_*$ induces a surjection $(\phi_q)_{*,c}: K(\mathbf{J}_c) \to K(\mathbf{H}_q)_c$. The surjection $(\phi_q)_{*,c}$ maps the $\mathbf{J}_c$-module E to the sum of simple constituents M of E_q with $c_M = c$, where E_q is the $\mathbf{H}_q$-module obtaining from E and the homomorphism $\phi_{q,c}: \mathbf{H}_q \to \mathbf{J} \to \mathbf{J}_c$.

For each $\mathbf{J}_c$-module E, we also denote $(\phi_q)_{*,c}(E)$ the direct sum of simple constituents of E_q to which the attached two-sided cell is c. When E is a simple

$\mathbf{J}_c$-module, I hope that $(\phi_q)_{*,c}(E)$ is either 0 or a simple $\mathbf{H}_q$-module. Furthermore, let E_1, E_2 be simple $\mathbf{J}_c$-modules such that $(\phi_q)_{*,c}(E_1) \neq 0$, I hope that $(\phi_q)_{*,c}(E_1) \simeq (\phi_q)_{*,c}(E_2)$ as $\mathbf{H}_q$-modules if and only if $E_1 \simeq E_2$ as $\mathbf{J}_c$-modules. If it is true, then the set $\{(\phi_q)_{*,c}(E) \mid c$ two-sided cell of W, E a simple $\mathbf{J}_c$-module (up to isomorphisms)$\}-\{0\}$ is a complete set of simple $\mathbf{H}_q$-modules, i.e. any simple $\mathbf{H}_q$-mdoule is isomorphic to some $(\phi_q)_{*,c}(E)$ and any two modules in the above set are not isomorphic. Then we get a classification of simple $\mathbf{H}_q$-modules. When q is not a root of 1 or $q = 1$, the above idea is valid (see [L13]).

In chapter 11 we apply the above idea to classify the simple $\mathbf{H}_q$-modules under the assumption that W is of type $\tilde{G}_2$ or $\tilde{B}_2$.

Lusztig conjectured that $(\phi_q)_*$ is an isomorphism if and only if $\sum_{w \in W_0} q^{l(w)}$ is not zero (see [L17]). When $(\phi_q)_*$ is an isomorphism, Lusztig showed that the set $\{(\phi_q)_{*,c}(E) \mid c$ two-sided cell of W, E a simple $\mathbf{J}_c$-module (up to isomorphisms)$\}$ is a complete set of simple $\mathbf{H}_q$-modules [L13]. Lusztig also showed that $(\phi_q)_*$ is an isomorphism when q is not a root of 1 or $q = 1$ (see [L13]). Note that $(\phi_q)_{*,c}$ ($c \in \mathrm{Cell}(W)$) is an isomorphism if $(\phi_q)_*$ is an isomorphism.

5.14. A conjecture of Lusztig concerned with the structure of $\mathbf{J}_c$. In this section we shall assume that G is a simply connected simple algebraic group. In [L14] Lusztig gives a nice conjecture concerned with the structure of the ring $\mathbf{J}_c$. Now we state his conjecture.

For each two-sided cell c of W, we have the corresponding nilpotent G-orbit $\mathcal{N}_c$ in $\mathbf{g}$. Choose an element $N \in \mathcal{N}_c$ and let F_c be a maximal reductive subgroup of $C_G(N)$. Lusztig conjectured that there exists a finite F_c-set Y and a bijection

$$\pi: c \xrightarrow{\sim} \text{ the set of irreducible } F_c\text{-v.b. on } Y \times Y(\text{ up to isomorphisms})$$

such that $t_w \to \pi(w)$ defines a $\mathbb{C}$-algebra isomorphism (preserving the unit element) between $\mathbf{J}_c$ and $\mathbf{K}_{F_c}(Y \times Y)$ and $\pi(w^{-1}) = \widetilde{\pi(w)}$, $w \in c$. When c is the lowest two-sided cell c_0, the conjecture was verified in [X2]. In chapter 11 we show that the conjecture is true when W is of type $\tilde{G}_2$, $\tilde{B}_2$. When W is of type $\tilde{A}_2$, $\tilde{A}_1$, we know the conjecture is valid (see [X3]).

Let c, N, F_c be as above. Note that $\mathcal{B}_N$ is an F_c-variety. Perhaps Lusztig's original ideal is the following conjecture.

5.15. Conjecture. *There exists a bijection $\pi: c \xrightarrow{\sim}$ the set of irreducible F_c-v.b. on $\mathcal{B}_N \times \mathcal{B}_N$ (up to isomorphisms) such that*

(i). *The map $t_w \to \pi(w)$ defines a $\mathbb{C}$-algebra isomorphism (preserving the unit element) between $\mathbf{J}_c$ and the convolution algebra $\mathbf{K}_{F_c}(\mathcal{B}_N \times \mathcal{B}_N)$.*

(ii). *For any w in c we have $\pi(w^{-1}) = \widetilde{\pi(w)}$.*

(iii). *The homomorphism ϕ in 2.7(c) has a natural geometric interpretation.*

5.16. Let G be a simply connected simple algebraic group over $\mathbb{C}$ and B a Borel subgroup of G. We shall identify G/B with $\mathcal{B}$. It is easy to describe the convolution algebra $\mathbf{K}_G(\mathcal{B} \times \mathcal{B})$.

We recall some steps in the proof of [KL4, Prop. 1.6]. Let T be a maximal torus in B and let $X = \mathrm{Hom}(T, \mathbb{C}^*)$ be the character group of T. Then we have $\mathbf{R}_T = \mathbb{C}[X]$ and $\mathbf{R}_G = \mathbb{C}[X]^{W_0}$. In 2.3 we have defined an element $x_w \in X$ for each w in W_0 and we know that $\mathbf{R}_T$ is a free $\mathbf{R}_G$-module with a basis $x_w, w \in W_0$ (see 2.3 (a)). Denote ρ the product of all fundamental weights in X. Then ρ^2 is the product of all positive roots in R.

Let $(,) : \mathbf{K}_G(\mathcal{B}) \times \mathbf{K}_G(\mathcal{B}) \to \mathbf{R}_G$ be the pairing defined by $(E, E') = \pi_*(E \otimes E')$ where $\pi : \mathcal{B} \to$ point is the natural map and π_* is direct image in $\mathbf{K}_G$-theory. Using the natural identification $\mathbf{K}_G(\mathcal{B}) = \mathbf{R}_B = \mathbf{R}_T$, this pairing becomes the pairing $(,) : \mathbf{R}_T \times \mathbf{R}_T \to \mathbf{R}_G$ given by

$$(x, y) = \sum_{w \in W_0} (-1)^{l(w)} w(xy\rho) / \sum_{w \in W_0} (-1)^{l(w)} w(\rho).$$

(Here we need Weyl's character formula.)

(a). We can find elements $y_u, u \in W_0$ such that $(x_w, y_u) = \delta_{w,u}$ for any w, u in W_0. Note that the elements $y_u, u \in W_0$ form an $\mathbf{R}_G$-basis of $\mathbf{R}_T$. (See the proof of [KL4, 1.6].)

According to [KL4, 1.6] and its proof, we have

(b). The external tensor product in $\mathbf{K}_G$-theory

$$\boxtimes : \mathbf{K}_G(\mathcal{B}) \otimes_{\mathbf{R}_G} \mathbf{K}_G(\mathcal{B}) \to \mathbf{K}_G(\mathcal{B} \times \mathcal{B})$$

is an isomorphism of $\mathbf{R}_G$-module.

(c). Its inverse is given by

$$\Phi : \mathbf{K}_G(\mathcal{B} \times \mathcal{B}) \to \mathbf{K}_G(\mathcal{B}) \otimes_{\mathbf{R}_G} \mathbf{K}_G(\mathcal{B})$$

$$\Phi(\xi) = \sum_{w \in W_0} y_w \otimes (\pi_2)_*(\pi_1^*(x_w) \otimes \xi),$$

where $\pi_1, \pi_2 : \mathcal{B} \times \mathcal{B} \to \mathcal{B}$ are two projections and π_1^* (resp. $(\pi_2)_*$) are inverse (resp. direct) image in $\mathbf{K}_G$-theory; here we regard x_w and y_w as elements in $\mathbf{K}_G(\mathcal{B}) = \mathbf{R}_T$.

(d). $\Phi(\eta_1 \boxtimes \eta_2) = \eta_1 \otimes \eta_2$, for any η_1, η_2 in $\mathbf{K}_G(\mathcal{B})$.

Based on (b), (c) and (d), we may get the following facts.

(e). The external tensor product in $\mathbf{K}_G$-theory

$$\boxtimes : \mathbf{K}_G(\mathcal{B}) \otimes_{\mathbf{R}_G} \mathbf{K}_G(\mathcal{B}) \otimes_{\mathbf{R}_G} \mathbf{K}_G(\mathcal{B}) \to \mathbf{K}_G(\mathcal{B} \times \mathcal{B} \times \mathcal{B})$$

is an isomorphism of $\mathbf{R}_G$-module.

(f). Its inverse is given by

$$\Psi : \mathbf{K}_G(\mathcal{B} \times \mathcal{B} \times \mathcal{B}) \xrightarrow{\Psi'} \mathbf{K}_G(\mathcal{B}) \otimes_{\mathbf{R}_G} \mathbf{K}_G(\mathcal{B} \times \mathcal{B})$$

$$\xrightarrow{\mathrm{id} \otimes \Phi} \mathbf{K}_G(\mathcal{B}) \otimes_{\mathbf{R}_G} \mathbf{K}_G(\mathcal{B}) \otimes_{\mathbf{R}_G} \mathbf{K}_G(\mathcal{B})$$

$$\Psi'(\xi) = \sum_{w \in W_0} y_w \otimes (p_{23})_*(p_1^*(x_w) \otimes \xi),$$

where $p_{23}, p_1 : \mathcal{B} \times \mathcal{B} \times \mathcal{B} \to \mathcal{B} \times \mathcal{B}$, $\mathcal{B}$ are two obvious projections and p_1^* (resp. $(p_{23})_*$) are inverse (resp. direct) image in $\mathbf{K}_G$-theory; here we regard x_w and y_w as elements in $\mathbf{K}_G(\mathcal{B}) = \mathbf{R}_T$.

(g). $\Psi(\eta_1 \boxtimes \eta_2 \boxtimes \eta_3) = \eta_1 \otimes \eta_2 \otimes \eta_3$, for any η_1, η_2, η_3 in $\mathbf{K}_G(\mathcal{B})$.

Using (b) we may identify $\mathbf{K}_G(\mathcal{B} \times \mathcal{B})$ with $\mathbf{K}_G(\mathcal{B}) \otimes_{\mathbf{R}_G} \mathbf{K}_G(\mathcal{B})$. Let $V = x_w\xi \boxtimes y_u$ and $V' = x_{u'} \boxtimes \eta y_v$ be two elements in $\mathbf{K}_G(\mathcal{B} \times \mathcal{B})$, where w, u, u' and v are elements in W_0, and ξ, η are elements in $\mathbf{R}_G$. Let $p_{12}, p_{23}, p_{13} : \mathcal{B} \times \mathcal{B} \times \mathcal{B} \to \mathcal{B} \times \mathcal{B}$ be the obvious projections. According to the definition of the convolution $*$ in 5.2, we have

$$V * V' = (Rp_{13})_*(p_{12}^* V \overset{L}{\underset{\mathcal{O}_{\mathcal{B} \times \mathcal{B} \times \mathcal{B}}}{\otimes}} p_{23}^* V')$$

(5.16.1)
$$= (p_{13})_*(x_w\xi \boxtimes y_u x_{u'} \boxtimes \eta y_v)$$
$$= \pi_*(y_u x_{u'})(x_w\xi \boxtimes \eta y_v)$$
$$= (x_{u'}, y_u)(x_w\xi \boxtimes \eta y_v)$$

Therefore the map $x_w\xi \boxtimes y_u \to P_{w,u}(\xi)$, $w, u \in W_0$, $\xi \in \mathbf{R}_G$, defines an isomorphism of $\mathbf{R}_G$-algebra between the convolution algebra $\mathbf{K}_G(\mathcal{B} \times \mathcal{B})$ and the $W_0 \times W_0$-matrix ring over $\mathbf{R}_G$, where $P_{w,u}(\xi)$ stands for the matrix whose entry is ξ at (w, u) and is zero otherwise. Let $i : \mathcal{B} \to \mathcal{B} \times \mathcal{B}$ be the diagonal embedding. Then $i_*(\mathbb{C}) = \sum_{w \in W_0} x_w \boxtimes y_w$ is the unit of the convolution algebra $\mathbf{K}_G(\mathcal{B} \times \mathcal{B})$ (cf. [KL4, 1.7]). It is known that $\mathbf{J}_{c_0}$ is isomorphic to the $W_0 \times W_0$-matrix ring over $\mathbf{R}_G$ (see [X2]).

Unfortunately it is difficult to classify the G-equivariant vector bundles on $\mathcal{B} \times \mathcal{B}$ (or equivalently to classify the B-equivariant vector bundles on G/B). I am grateful to Professor G. Lusztig for explaining me this.

Affine Hecke algebras with two parameters

5.17. In the rest part of the chapter we shall consider the representations of affine Hecke algebras of two parameters. We mainly follow the line in [L9]. We shall assume that G is a simple, simply connected algebraic group and keep the notations in 2.8-2.9.

Let $\tilde{G} = G \times \mathbb{C}^* \times \mathbb{C}^*$, then we may define a $\tilde{G}$-action on $\mathcal{B}$ as follows: G acts on $\mathcal{B}$ through adjoint action and $\mathbb{C}^* \times \mathbb{C}^*$ acts on $\mathcal{B}$ trivially. We have

$$K_{\tilde{G}}(\mathcal{B}) = K_G(\mathcal{B}) \otimes R_{\mathbb{C}^* \times \mathbb{C}^*} = K_G(\mathcal{B}) \otimes \mathbb{Z}[\mathbf{u}^{\pm 2}, \mathbf{v}^{\pm 2}],$$

where $\mathbf{u}^2, \mathbf{v}^2$ are the generators of $R_{\mathbb{C}^* \times \mathbb{C}^*}$ corresponding to the obvious projections: $p_1, p_2 : \mathbb{C}^* \times \mathbb{C}^* \to \mathbb{C}^*$, and $R_{\mathbb{C}^* \times \mathbb{C}^*} = K_{\mathbb{C}^* \times \mathbb{C}^*}(point)$ is the Grothendieck group of the rational representations of $\mathbb{C}^* \times \mathbb{C}^*$.

For each $s \in S_0$, we denote $\mathcal{P}_s$ the variety of all parabolic subalgebras of $\mathbf{g}$ corresponding to s and let $\pi_s : \mathcal{B} \to \mathcal{P}_s$ be the natural map. There is a unique endomorphism

(5.17.1)
$$T_s : K_{\tilde{G}}(\mathcal{B}) \to K_{\tilde{G}}(\mathcal{B})$$

with the following property: if E is a $\tilde{G}$-equivariant algebraic vector bundle on $\mathcal{B}$, then

$$(5.17.2) \qquad E + T_s E = (\pi_s^*(\pi_s)_*(E^*) - \pi_s^*(\pi_s)_*(E^* \otimes \Omega_s^1))^*,$$

where Ω_s^1 is the line bundle on $\mathcal{B}$ of holomorphic differential 1-forms along the fibres of π_s, G acts on Ω_s^1 in an obvious way, let $\mathbb{C}^*$ act on each fibre of Ω_s^1 by scalar multiplication, and then $\mathbb{C}^* \times \mathbb{C}^*$ acts on each fibre of Ω_s^1 through the projection p_1 if $s \in S_0'$ and through the projection p_2 if $s \in S_0''$. Here $(\pi_s)_*(E^*)$ is the alternating sum of the higher direct images of E^* under π_s in the category of coherent sheaves (we use E'^* for the dual of a vector bundle E'); these higher direct images are again $\tilde{G}$-equivariant algebraic vector bundle on $\mathcal{P}_s$ (see [T]), hence their alternating sums defines an element in $K_{\tilde{G}}(\mathcal{P}_s)$.

For any element $x \in X$, we define an endomorphism

$$(5.17.3) \qquad\qquad \theta_x : K_{\tilde{G}}(\mathcal{B}) \to K_{\tilde{G}}(\mathcal{B})$$

by

$$(5.17.4) \qquad\qquad \theta_x E = E \otimes L_x,$$

where L_x is the line bundle on $\mathcal{B}$ associated to the weight $x : T \to \mathbb{C}^*$, it is a $\tilde{G}$-equivariant bundle with the obvious action of G and with trivial action of $\mathbb{C}^* \times \mathbb{C}^*$.

5.18. Proposition. *The endomorphisms T_s and θ_x of $K_{\tilde{G}}(\mathcal{B})$ defined in 5.17 give rise to a left $\tilde{H}^\circ$-module structure on $K_{\tilde{G}}(\mathcal{B})$. (The action of $\mathbb{Z}[\mathbf{u}^{\pm 2}, \mathbf{v}^{\pm 2}] \subset \tilde{H}^\circ$ is defined to be the same as the restriction to $R_{\mathbb{C}^* \times \mathbb{C}^*}$ of the action of $R_{\tilde{G}}$, note that $K_{\tilde{G}}(\mathcal{B})$ is naturally an $R_{\tilde{G}}$-module.) This $\tilde{H}^\circ$-module structure commutes with the $R_{\tilde{G}}$-module structure on $K_{\tilde{G}}(\mathcal{B})$.*

Proof. We identify $K_{\tilde{G}}(\mathcal{B})$ with $B'[X]$ and identify $R_{\tilde{G}}$ with $B'[X]^{W_0}$, where $B' = \mathbb{Z}[\mathbf{u}^{\pm 2}, \mathbf{v}^{\pm 2}]$. Then the canonical ring homomorphism $R_{\tilde{G}} \to K_{\tilde{G}}(\mathcal{B})$ becomes the inclusion $B'[X]^{W_0} \hookrightarrow B'[X]$. Under these identifications, the endomorphisms in 5.17 become B'-linear maps and satisfy

$$(5.18.1) \qquad T_s(x) = \frac{s(x)\alpha_s - x\alpha_s}{\alpha_s - 1} + \mathbf{u}^2 \frac{x\alpha_s - s(x)}{\alpha_s - 1}, \quad s \in S_0', \quad x \in X,$$

$$(5.18.2) \qquad T_s(x) = \frac{s(x)\alpha_s - x\alpha_s}{\alpha_s - 1} + \mathbf{v}^2 \frac{x\alpha_s - s(x)}{\alpha_s - 1}, \quad s \in S_0'', \quad x \in X,$$

$$(5.18.3) \qquad\qquad \theta_y(x) = xy.$$

Let $\mathcal{I}$ be the left ideal of $\tilde{H}^\circ$ generated by $C = \sum_{w \in W_0} T_w$, then $\theta_x C$, $x \in X$ form a B'-basis of $\mathcal{I}$. Under the natural map $B'[X] \to \mathcal{I}$, $x \to \theta_x C$, the actions (5.18.1-2) (resp. (5.18.3) become left multiplications by T_s (resp. θ_x) for the left $\tilde{H}^\circ$-module structure of the left ideal $\mathcal{I}$. The proposition is proved.

5.19. Motivated by a conjecture in [L4] we formulate a conjecture, which is an analogue of the $(*)$ in the introduction of the book. We need some notations.

We set

$$(5.19.1) \qquad f_{W_0}(\mathbf{u}^2, \mathbf{v}^2) = \sum_{w \in W_0} \mathbf{u}^{2l'(w)} \mathbf{v}^{2l''(w)}.$$

Then
(5.19.2)
$$f_{W_0}(\mathbf{u}^2, \mathbf{v}^2) = \prod_{i=1}^{n}(1 + \mathbf{u}^2 + \cdots + \mathbf{u}^{2(i-1)})(1 + \mathbf{u}^{2(i-1)}\mathbf{v}^2), \quad \text{if } G \text{ is of type } B_n \text{ or } C_n,$$

$$(5.19.3) \qquad f_{W_0}(\mathbf{u}^2, \mathbf{v}^2) = (1 + \mathbf{u}^2)(1 + \mathbf{u}^2 + \mathbf{u}^4)(1 + \mathbf{v}^2)(1 + \mathbf{v}^2 + \mathbf{v}^4) \cdot$$

$$(1 + \mathbf{u}^2\mathbf{v}^4)(1 + \mathbf{u}^4\mathbf{v}^2)(1 + \mathbf{u}^2\mathbf{v}^2)(1 + \mathbf{u}^4\mathbf{v}^4)(1 + \mathbf{u}^6\mathbf{v}^6), \quad \text{if } G \text{ is of type } F_4,$$

$$(5.19.4) \quad f_{W_0}(\mathbf{u}^2, \mathbf{v}^2) = (1 + \mathbf{u}^2)(1 + \mathbf{v}^2)(1 + \mathbf{u}^2\mathbf{v}^2 + \mathbf{u}^4\mathbf{v}^4), \quad \text{if } G \text{ is of type } G_2,$$

Let T be a maxiaml torus of G and $\mathbf{t}$ its Lie algebra. We have

$$(5.19.5) \qquad \mathbf{g} = \mathbf{t} \oplus (\bigoplus_{\alpha \in R} \mathbf{g}_\alpha).$$

For any $s \in T$ and $(a, b) \in \mathbb{C}^* \times \mathbb{C}^*$, we set

$$(5.19.6) \qquad \mathbf{g}_{s,a,b} = \{0\} \cup ((\bigoplus_{\substack{\alpha \in R' \\ \alpha(s)=a}} \mathbf{g}_\alpha) \oplus (\bigoplus_{\substack{\alpha \in R'' \\ \alpha(s)=b}} \mathbf{g}_\alpha)),$$

$$(5.19.7) \qquad \mathcal{N}_{s,a,b} = \mathcal{N} \cap \mathbf{g}_{s,a,b}.$$

For each $N \in \mathcal{N}_{s,a,b}$, let $A(s, N)$ and $A(s, N)^\vee$ be as in 5.6.

5.20. Conjecture. If $f_{W_0}(a, b) \neq 0$, then there is a natural one-to-one correspondence between the set of isomorphism classes of simple $\tilde{H}_{a,b}$-modules and the set of G-conjugacy classes of the triples (s, N, ρ), where $s \in T$, $N \in \mathcal{N}_{s,a,b}$, and $\rho \in A(s, N)^\vee$. When $f_{W_0}(a, b) = 0$, no such natural correspondence exists.

6. An Equivalence Relation in $T \times \mathbb{C}^*$

Let G be a connected reductive group over $\mathbb{C}$ and T a maximal torus of G. Motivated by the result 5.6(e) (due to Ginzburg) we introduce an equivalence relation in the set $T \times \mathbb{C}^*$. We establish some properties of the equivalence relation. Combining these properties, 5.6(d) (due to Kazhdan and Lusztig) and 5.6(e) we can prove that the conjecture $(*)$ in the introduction of the book is true when the order of q is not too small. The main results are Theorem 6.5 and Theorem 6.6. For type A_n our results also confirm a conjecture of Zelevinsky [Z, 8.7].

6.1. The equivalence relation. Recall that $\mathbf{g}$ denotes the Lie algebra of G and $\mathcal{N}$ is the set of all nilpotent elements in $\mathbf{g}$. For each element (s,q) in $T \times \mathbb{C}^*$ we have defined the sets (cf. 5.6)

$$\mathbf{g}_{s,q} = \{\xi \in \mathbf{g} \mid s.\xi = q\xi\},$$

$$\mathcal{N}_{s,q} = \{\xi \in \mathcal{N} \mid s.\xi = q\xi\}.$$

For any two elements (s,q) and (t,r) in $T \times \mathbb{C}^*$ we shall write $(s,q) \sim (t,r)$ if $\mathcal{N}_{s,q} = \mathcal{N}_{t,r}$ and $C_G(s) = C_G(t)$. Obviously the relation $\sim$ in $T \times \mathbb{C}^*$ is an equivalence relation.

We now discuss the conditions for $(s,q) \sim (t,r)$. Denote $\mathbf{t}$ the Lie algebra of T and $R \subset \text{Hom}(T, \mathbb{C}^*)$ the root system associated to G. For each $\alpha \in R$ we denote $\mathbf{g}_\alpha$ the corresponding root subspace of $\mathbf{g}$ and G_α the corresponding one-parameter subgroup of G. Then $\mathbf{g} = \mathbf{t} \oplus (\bigoplus_{\alpha \in R} \mathbf{g}_\alpha)$. Let (s,q) and (t,r) be elements in $T \times \mathbb{C}^*$. We have

(a). Assume that both q and r are not 1, or both q and r are equal to 1, then $\mathcal{N}_{s,q} = \mathcal{N}_{t,r}$ if and only if $\mathbf{g}_{s,q} = \mathbf{g}_{t,r}$.

(b). Assume that both q and r are not 1, or both q and r are equal to 1, then $\mathcal{N}_{s,q} = \mathcal{N}_{t,r}$ if and only if $R_{s,q} = R_{t,r}$, where $R_{s,q} := \{\alpha \in R \mid \alpha(s) = q\}$ and $R_{t,r}$ is defined similarly.

(c). Assume that $q = 1$ and $r \neq 1$, then $\mathcal{N}_{s,q} = \mathcal{N}_{t,r}$ if and only if both $R_{s,q}$ and $R_{t,r}$ are empty, i.e., if and only if $\mathcal{N}_{s,q} = \mathcal{N}_{t,r} = \{0\}$.

Let $W_0 = N_G(T)/T$ be the Weyl group of G. The following facts (d) and (e) are well known (cf. [St1]).

(d). $C_G(s)$ is generated by T, G_α ($\alpha \in R$ and $\alpha(s) = 1$) and elements $\dot{w} \in N_G(T)$ satisfying $\dot{w}s = s\dot{w}$.

(e). Assume that G has a simply connected derived group, then $C_G(s)$ is generated by T, G_α ($\alpha \in R$ and $\alpha(s) = 1$).

By (e) we get

(f). Assume that G has a simply connected derived group, then $C_G(s) = C_G(t)$ if and only if $R_{s,1} = R_{t,1}$.

From (b) and (d) we obtain

(g). The number of equivalence classes in $T \times \mathbb{C}^*$ with respect to $\sim$ is finite.

Actually it is easy to see that the number of the equivalence classes is less than $2^{2|R|+|W_0|+1}$.

Let $\varphi : G \to G'$ be a surjective homomorphism from G to an algebraic group G' (over $\mathbb{C}$) such that its kernel is contained in the center of G. Then $T' = \varphi(T)$ is a maximal torus of G and we may identify the root system of G with the root system of G' through the natural embedding $\mathrm{Hom}(T', \mathbb{C}^*) \hookrightarrow \mathrm{Hom}(T, \mathbb{C}^*)$. Let $u \in T$. Then u is in the center of G if and only if $\alpha(u) = 1$ for all roots α in R. Therefore we have

(h). Assume $(s, q), (t, r) \in T \times \mathbb{C}^*$, then $R_{s,q} = R_{t,r}$ if and only if $R_{\varphi(s),q} = R_{\varphi(t),r}$

In (j) and (k), G is assumed to have a simply connected derived group; both q and r are not 1, or both q and r are equal to 1; $\varphi : G \to G'$ is as above; and $(s, q), (t, r)$ are elements in $T \times \mathbb{C}^*$.

(j). If $(\varphi(s), q) \sim (\varphi(t), r)$, then $(s, q) \sim (t, r)$.

(k). If $R_{\varphi(s),q} = R_{\varphi(t),r}$ and $R_{\varphi(s),1} = R_{\varphi(t),1}$, then $(s, q) \sim (t, r)$.

Remark. It is possible that $R_{s,q} = R_{t,r}$ but $\mathbf{g}_{s,q} \neq \mathbf{g}_{t,r}$ and $\mathcal{N}_{s,q} \neq \mathcal{N}_{t,r}$. As an example we consider $G = SL_2(\mathbb{C})$. Let $T = \{\mathrm{diag}(a, a^{-1}) \mid a \in \mathbb{C}^*\}$ and let $\alpha : T \to \mathbb{C}^*$ be defined by $\alpha : \mathrm{diag}(a, a^{-1}) \to a^2$. Then $R = \{\alpha, -\alpha\}$. Let $s = \mathrm{diag}(-1, -1)$, $q = 1$; and $t = \mathrm{diag}(\sqrt{-1}, -\sqrt{-1})$, $r = -1$. Then $R_{s,q} = R_{t,r} = R$, but

$$\mathbf{g}_{s,q} = \mathbf{g} \neq \mathbf{g}_{t,r} = \mathbf{g}_\alpha \oplus \mathbf{g}_{-\alpha},$$
$$\mathcal{N}_{s,q} = \mathcal{N} \neq \mathcal{N}_{t,r} = \mathbf{g}_\alpha \cup \mathbf{g}_{-\alpha}.$$

6.2. Proposition. *Assume that G has a simply connected derived group. Let (s, q) and (t, r) be two elements in $T \times \mathbb{C}^*$ such that $(s, q) \sim (t, r)$. Then*

(i). *We have $\mathcal{B}^s = \mathcal{B}^t$ and $Z^{s,q} = Z^{t,r}$. (See 5.6 (e) for the definition of $Z^{s,q}$.)*

(ii). *For any nilpotent element N in $\mathcal{N}_{s,q} (= \mathcal{N}_{t,r})$, we have $\mathcal{B}_N^s = \mathcal{B}_N^t$.*

(iii). *We have $A(s, N) = A(t, N)$ and $A(s, N)^\vee = A(t, N)^\vee$.*

Proof. (i). Since $C_G(s) = C_G(t)$, according to Steinberg (see [St1]) we have $\mathcal{B}^s = \mathcal{B}^t$. In addition we have $Z^{s,q} = Z^{t,r}$ since $\mathcal{N}_{s,q} = \mathcal{N}_{t,r}$. (ii) follows from $\mathcal{B}_N^s = \mathcal{B}^s \cap \mathcal{B}_N$, $\mathcal{B}_N^t = \mathcal{B}^t \cap \mathcal{B}_N$. And the third assertion is obvious by (ii) and the definition of $\sim$.

6.3. Proposition. *Keep the set up and notations in 5.5-5.6 and in 6.2. Assume that $(s, q) \sim (t, r)$, then*

(i). *The $\mathbb{C}$-algebras $\mathbf{H}_{s,q}$ and $\mathbf{H}_{t,r}$ are isomorphic.*

(ii). *For any $\rho \in A(s, N)^\vee = A(t, N)^\vee$, the standard module $M_{s,N,q,\rho}$ has a unique quotient if and only if $M_{t,N,r,\rho}$ has a unique quotient.*

(iii). *Let N' be another nilpotent element in $\mathcal{N}_{s,q}(= \mathcal{N}_{t,r})$ and $\rho' \in A(s,N')^\vee$. Assume that the standard modules $M_{s,N,q,\rho}$, $M_{t,N,r,\rho}$, $M_{s,N',q,\rho'}$ and $M_{t,N',r,\rho'}$ all possess a unique simple quotient module respectively, denote by $L_{s,N,q,\rho}$, $L_{t,N,r,\rho}$, $L_{s,N',q,\rho'}$ and $L_{t,N',r,\rho'}$, respectively. then $L_{s,N,q,\rho}$ is isomorphic to $L_{s,N',q,\rho'}$ if and only if $L_{t,N,r,\rho}$ is isomorphic to $L_{t,N',r,\rho'}$.*

Proof. The assertion (i) follows from 5.6(e) and 6.2 (i). The other assertions follow from these facts: the definitions of standard modules, (i), 5.6(d) and 6.2.

The proposition is proved.

6.4. For simplicity in the rest part of this chapter we shall assume that G is simply connected and simple (over $\mathbb{C}$) except specified indications. Let

(6.6.1) $$\mathbb{C}^*_{W_0} = \{q \in \mathbb{C}^* \mid \text{the order of } q > e_n + 1\},$$

where e_n is the maximal exponent of W_0. We shall write $o(q)$ for the order of q.

6.5. Theorem. *Let G be a simply connected, simple algebraic group over $\mathbb{C}$.*

(i). *For any (s,q) in $T \times \mathbb{C}^*_{W_0}$ and r in $\mathbb{C}^*_{W_0}$, there exists t in T such that $(s,q) \sim (t,r)$. In particular we have*

(ii). *For any element (s,q) in $T \times \mathbb{C}^*_{W_0}$, there exists (t,r) in $T \times \mathbb{C}^*_{W_0}$ such that r is not a root of 1 and $(s,q) \sim (t,r)$.*

We shall prove the theorem case by case. Combining 6.3, 6.5 and 5.6(d) we get

6.6. Theorem. *Let G be a simply connected, simple algebraic group over $\mathbb{C}$. Assume that $o(q) > e_n + 1$, then*

(i). *Each standard module $M_{s,N,q,\rho}$ has a unique quotient module, denote by $L_{s,N,q,\rho}$.*

(ii). *The $\mathbf{H}_q$-modules $L_{s,N,q,\rho}$ and $L_{s',N',q,\rho'}$ are isomorphic if and only if (s,N,q,ρ) and (s',N',q,ρ') are G-conjugate.*

(iii). *Each simple $\mathbf{H}_q$-module is isomorphic to some $L_{s,N,q,\rho}$.*

6.7. We call an element $(s,q) \in T \times \mathbb{C}^*$ is good if there exists an element (t,r) in $T \times \mathbb{C}^*_{W_0}$ such that $(s,q) \sim (t,r)$, is semi-good if $(s,q) \sim (t,r)$ for some $(t,r) \in T \times \mathbb{C}^*$ with $f_{W_0}(r) \neq 0$, is bad if (s,q) is not semi-good, where

$$f_{W_0} = f_{W_0}(\mathbf{q}) = \sum_{w \in W_0} \mathbf{q}^{l(w)} \in \mathbf{A}.$$

Example: Suppose that G is simple and of rank n. Let $s \in T$ be such that $\alpha(s) = q$ for all $\alpha \in \Delta$ (set of the simple roots in R) and assume that $o(q) = e_n$, the biggest exponent of W_0. If G is not of type A_n, then $e_n - 1$ is not exponent of W_0, and (s,q) is semi-good but not good.

We also prove the following result through case by case analysis. The result will be needed in next chapter.

6.8. Theorem. *Let $(s, q) \in T \times \mathbb{C}^*$, then (s, q) is bad if and only if $\mathbf{g}_{s,q}$ is not equal to $\mathcal{N}_{s,q}$.*

6.9. Even if (s, q) is bad, the variety $\mathcal{N}_{s,q}$ is possible to be irreducible. For example, let $G = SL_3(\mathbb{C})$, and $s = \mathrm{diag}(-1, 1 - 1) \in G$, then $(s, -1)$ is bad but $\mathcal{N}_{s,-1}$ is irreducible (cf. [KL4, 5.15] and 7.8 in next chapter).

We shall need a result of Lusztig in [L17], which can be verified directly when G is a classical group. For each $w \in W_0$, choose an element $\dot{w} \in N_G(T)$ such that its image in W_0 is w. Note that $\mathbf{t}$ (the Lie algebra of T) is W_0-stable. Let $f_w(\mathbf{q}) = det(\mathbf{q} - w)$ be the eigenpolynomial of w on the space $\mathbf{t}$.

(a) Every element g in $\dot{w}T$ is semisimple.

Proof. Consider the adjoint representation Ad: $G \to GL(\mathbf{g})$. It is easy to check that the image $\mathrm{Ad}(g)$ of g is semisimple. It is known that the kernel of Ad is the center of G. So g is semisimple.

A result of Lusztig in [L17] can be expressed as

(b) Assume that $q \neq 1$, then $\mathbf{g}_{s,q}$ is not equal to $\mathcal{N}_{s,q}$ if and only if s is conjugate to some element in $\dot{w}T$ such that $f_w(q) = 0$.

We need a few more notations: $\mathbf{g}^+$ stands for $\underset{\alpha \in R^+}{\oplus} \mathbf{g}_\alpha$, and T_{reg} denotes the set of regular elements in the maximal torus T. Now we begin our proofs of 6.5 and of 6.8 through case by case analysis. We will need the facts 6.1 (j) and 6.1 (k).

6.10. Type A_n: We have $G = SL_{n+1}(\mathbb{C})$. We choose the maximal torus T to be the set of the diagonal matrices in G.

Let $\alpha_{ij}, x_k \in X = \mathrm{Hom}(T, \mathbb{C}^*)$, $1 \leq i < j \leq n+1$, $1 \leq k \leq n$, be defined as follows:

$$\alpha_{ij} : \mathrm{diag}\,(a_1, a_2, \cdots, a_{n+1}) \to a_i a_j^{-1},$$

$$x_k : \mathrm{diag}\,(a_1, a_2, \cdots, a_{n+1}) \to a_1 a_2 \cdots a_k.$$

Let $R^+ = \{\alpha_{ij} \mid 1 \leq i < j \leq n+1\}$. Then the set of simple roots is $\Delta = \{\alpha_{i,i+1} \mid 1 \leq i \leq n\}$, and $x_1, x_2, ..., x_n$ are the fundamental weights.

The normalizer $N_G(T)$ of T in G is generated by T and elements $P_i(-1)P_{ij}$ ($1 \leq i \neq j \leq n+1$), where $P_i(-1)$ is the matrix obtained by multiplying the i-th row of the identity matrix I_{n+1} with -1, and P_{ij} is the matrix obtained by exchanging the i-th row and the j-th row of the matrix I_{n+1}. The Weyl group $W_0 = N_G(T)/T$ is isomorphic to the symmetric group $\mathfrak{S}_{n+1}$ of degree $n+1$.

Let (s, q) be an element in $T \times \mathbb{C}^*$ and assume that $q \neq 1$. Obviously through an element of the Weyl group W_0, s is conjugate to certain element $D \in T$ of the following form:

(6.10.1) $$D = \mathrm{diag}(D_1, D_2, ..., D_k),$$

where

$$D_i = \begin{pmatrix} d_i q^{m_i} I_{r_{i,i}} & 0 & \cdots & 0 & 0 \\ 0 & d_i q^{m_i-1} I_{r_{i,i-1}} & \cdots & 0 & 0 \\ \vdots & \vdots & \ddots & \vdots & \vdots \\ 0 & 0 & \cdots & d_i q I_{r_{i,1}} & 0 \\ 0 & 0 & \cdots & 0 & d_i I_{r_{i,0}} \end{pmatrix}, \qquad 1 \le i \le k,$$

all $r_{i,j}$ are positive integers, and all m_i are non-negative integers, $d_i \in \mathbb{C}^*$, moreover, $\max\{m_1, m_2, ..., m_k\} < o(q)$, and

$$d_i q^m (d_j q^{m'})^{-1} \ne 1, q, q^{-1}, \quad \text{for any } 1 \le i \ne j \le k, \ 0 \le m \le m_i, \ 0 \le m' \le m_j.$$

Let (s, q) be as above. We have

(a). $\mathbf{g}_{s,q}$ is not equal to $\mathcal{N}_{s,q}$ if and only if $m_i + 1 = o(q)$ for some integer i in $[1, k]$.

(b). If $\mathbf{g}_{s,q} = \mathcal{N}_{s,q}$, then for an arbitrary $r \in \mathbb{C}^*$ with $o(r) - 1 > \max\{m_1, m_2, ..., m_k\}$, we can find some t in T such that $(s, q) \sim (t, r)$.

Proof. It is harmless to assume that $s = D$ (notations as above).

(a). If $o(q) - 1 > m_i$ for $i = 1, 2, ..., k$, then we have $\mathbf{g}_{s,q} \subset \mathbf{g}^+$, so $\mathbf{g}_{s,q} = \mathcal{N}_{s,q}$. Suppose that $m_i + 1 = o(q)$ for some i, then the exist positive roots $\beta_1, \beta_2, ..., \beta_{m_i}$ such that $\beta_1 + \cdots + \beta_j \in R^+$ for $j = 1, 2, ..., m_i$ and

$$\mathbf{g}'_{s,q} = \mathbf{g}_{\beta_1} + \mathbf{g}_{\beta_2} + \cdots + \mathbf{g}_{\beta_m} + \mathbf{g}_{-\beta_1-\beta_2-\cdots-\beta_m} \subseteq \mathbf{g}_{s,q},$$

where $m = m_i$. Note that $\mathbf{g}'_{s,q}$ contains semisimple elements, so $\mathbf{g}_{s,q}$ is not equal to $\mathcal{N}_{s,q}$.

(b). Assume that $o(q) - 1 > \max\{m_1, m_2, ..., m_k\}$. Choose nonzero complex numbers $a_1, ..., a_k \in \mathbb{C}^*$ such that $|a_i r^m (a_j r^{m'})^{-1}| > \max\{1, |r|, |r|^{-1}\}$ for arbitrary $1 \le i < j \le k$, $0 \le m \le m_i$, $0 \le m' \le m_j$, and such that $E = \mathrm{diag}(E_1, E_2, ..., E_k) \in G$, where

$$E_i = \begin{pmatrix} a_i r^{m_i} I_{r_{i,i}} & 0 & \cdots & 0 & 0 \\ 0 & a_i r^{m_i-1} I_{r_{i,i-1}} & \cdots & 0 & 0 \\ \vdots & \vdots & \ddots & \vdots & \vdots \\ 0 & 0 & \cdots & a_i r I_{r_{i,1}} & 0 \\ 0 & 0 & \cdots & 0 & a_i I_{r_{i,0}} \end{pmatrix}, \qquad 1 \le i \le k.$$

Then we have $(D, q) \sim (E, r)$.

The assertions (a-b) are proved.

(c). Let s be an element in T and q a primitive $(n+1)$-th root of 1. We have

(i). $\mathbf{g}_{s,q}$ is not equal to $\mathcal{N}_{s,q}$ if and only if s is conjugate to the element

$$\mathrm{diag}(q^{\frac{n}{2}}, q^{\frac{n-3}{2}}, ..., q^{\frac{3-n}{2}}, q^{-\frac{n}{2}}).$$

(ii). If $\mathbf{g}_{s,q} = \mathcal{N}_{s,q}$, then for any r in $\mathbb{C}^*_{W_0}$ we can find $t \in T$ such that $(s,q) \sim (t,r)$.

Proof. (i) follows from (a), and (ii) follows from (b).

(d). Let $(s,q) \in T \times \mathbb{C}^*$ with $q \neq 1$. Assume that $\mathbf{g}_{s,q}$ is not equal to $\mathcal{N}_{s,q}$, then there exists a sequence $t_1, t_2, ..., t_k, ...$ in T_{reg} such that

$$\mathbf{g}_{t_k,q} \neq \mathcal{N}_{t_k,q} \quad \text{and} \quad \lim_{k \to \infty} t_k = s.$$

(In this book all limits are with respect to the complex topology.)

Proof. By the proof of (a) we see that s is conjugate to certain element $D = \text{diag}(dq^m, dq^{m-1}, ..., dq, d, a_1, a_2, ..., a_{n-m}) \in T$, where $m+1$ is the order of q. Note that $q^{i-j} \neq 1$ for any different integers i,j in $[0,m]$. Choose positive numbers $b_1, b_2, ..., b_k, ...$, in the interval $(1, +\infty)$ such that $\lim_{k \to \infty} b_k = 1$ and such that

$$a_i b_k^{n-m-2i+1} a_j^{-1} b_k^{2j-1-n+m} \neq 1, \quad \text{for different integers } i,j \text{ in } [1, n-m];$$

$$a_i b_k^{n-m-2i+1} d^{-1} q^{-l} \neq 1, \quad \text{for } l = 0, 1, ..., m.$$

Let

$$t_k = \text{diag}(dq^m, dq^{m-1}, ..., dq, d, a_1 b_k^{n-m-1}, a_2 b_k^{n-m-3}, ..., a_{n-m} b_k^{m-n+1}) \in T,$$

then the sequence $t_1, t_2, ..., t_k, ...$ satisfies our requirement. The assertion is proved.

6.11. Type B_n. Since $\alpha(s) = 1$ for all $\alpha \in R$ whenever s is in the center of G, we may consider the special orthogonal group $SO_{2n+1}(\mathbb{C})$ instead of the spin group $Spin_{2n+1}(\mathbb{C})$. But the results are also valid for $Spin_{2n+1}(\mathbb{C})$ for the above reason (see also 6.1(j-k)).

The group is

$$G = SO_{2n+1}(\mathbb{C}) = \{g \in SL_{2n+1}(\mathbb{C}) \mid \tilde{g} \begin{pmatrix} 1 & 0 & 0 \\ 0 & 0 & I_n \\ 0 & I_n & 0 \end{pmatrix} g = \begin{pmatrix} 1 & 0 & 0 \\ 0 & 0 & I_n \\ 0 & I_n & 0 \end{pmatrix} \},$$

where $\tilde{g}$ is the transpose of g. We choose the maximal torus T to be the set of the diagonal matrices in G. Then

$$T = \{\text{diag}(1, a_1, a_2, ..., a_n, a_1^{-1}, a_2^{-1}, ..., a_n^{-1}) \mid a_1, a_2, ..., a_n \in \mathbb{C}^*\}.$$

The normalizer $N_G(T)$ of T in G is generated by T and elements $P_{ij} P_{i+n,j+n}$, $P_1(-1) P_{i,i+n}$ ($2 \leq i \neq j \leq n+1$), where P_{ij} is the matrix obtained by exchanging the i-th row and the j-th row of the matrix I_{2n+1}, and $P_1(-1)$ is the matrix obtained by multiplying the 1-st row of the matrix I_{2n+1} with -1. The Weyl group $W_0 = N_G(T)/T$ is isomorphic to the semi-direct product $(\mathbb{Z}/2\mathbb{Z})^n \ltimes \mathfrak{S}_n$.

Let $\alpha_{ij}, \beta_{ij}, \gamma_k \in X = \text{Hom}(T, \mathbb{C}^*)$, $1 \leq i < j \leq n, 1 \leq k \leq n$, be defined as follows:

$$\alpha_{ij} : \text{diag}(1, a_1, a_2, ..., a_n, a_1^{-1}, a_2^{-1}, ..., a_n^{-1}) \to a_i a_j^{-1},$$

$$\beta_{ij} : \text{diag}\,(1, a_1, a_2, ..., a_n, a_1^{-1}, a_2^{-1}, ..., a_n^{-1}) \rightarrow a_i a_j,$$

$$\gamma_k : \text{diag}\,(1, a_1, a_2, ..., a_n, a_1^{-1}, a_2^{-1}, ..., a_n^{-1}) \rightarrow a_k.$$

Let $R^+ = \{\alpha_{ij},\ \beta_{ij},\ \gamma_k \mid 1 \le i < j \le n,\ 1 \le k \le n\}$. Then the set of simple roots is $\Delta = \{\alpha_{i,i+1},\ \gamma_n \mid 1 \le i \le n-1\}$.

Let (s, q) be an element in $T \times \mathbb{C}^*$ and assume that $q \ne 1$. Obviously through an element of the Weyl group W_0, s is conjugate to certain element $D \in T$ of the following form:

(6.11.1) $$D = \text{diag}(1, D_1, D_2, ..., D_k, D_1^{-1}, D_2^{-1}, ..., D_k^{-1}),$$

where

$$D_i = \begin{pmatrix} d_i q^{m_i} I_{r_{i,i}} & 0 & \cdots & 0 & 0 \\ 0 & d_i q^{m_i - 1} I_{r_{i,i-1}} & \cdots & 0 & 0 \\ \vdots & \vdots & \ddots & \vdots & \vdots \\ 0 & 0 & \cdots & d_i q I_{r_{i,1}} & 0 \\ 0 & 0 & \cdots & 0 & d_i I_{r_{i,0}} \end{pmatrix}, \qquad 1 \le i \le k,$$

all $r_{i,j}$ are positive integers, and all m_i are non-negative integers, $d_i \in \mathbb{C}^*$, moreover $\max\{m_1, m_2, ..., m_k\} < o(q)$, and

$$d_i q^m (d_j q^{m'})^{\pm 1} \ne 1, q, q^{-1}, \quad \text{for any } 1 \le i \ne j \le k,\ 0 \le m \le m_i,\ 0 \le m' \le m_j.$$

We have

(a). $\mathbf{g}_{s,q}$ is not equal to $\mathcal{N}_{s,q}$ if and only if at least one of the following conditions is satisfied.

(i). There is some integer i in $[1, k]$ such that $m_i + 1 = o(q)$.

(ii). $o(q)$ is even and there are some integer i in $[1, k]$ and integer m in $[0, m_i]$ such that $d_i q^m = q$ and $2m_i - 2m + 2 \ge o(q)$.

Proof. We may prove the assertions as the case of type A_n.

(b). If $\mathbf{g}_{s,q}$ is not equal to $\mathcal{N}_{s,q}$, then there exists a sequence $t_1, t_2, ..., t_k, ...$ in T_{reg} such that

$$\mathbf{g}_{t_k,q} \ne \mathcal{N}_{t_k,q}, \quad \text{and} \quad \lim_{k \to \infty} t_k = s.$$

Proof. Assume that $\mathbf{g}_{s,q}$ is not equal to $\mathcal{N}_{s,q}$. By (a) we see that s is conjugate to certain element $D = \text{diag}(1, D_1, D_2, D_1^{-1}, D_2^{-1}) \in T$, such that

$$D_1 = \text{diag}(dq^m, dq^{m-1}, ..., dq, d), \qquad \text{for some } d \in \mathbb{C}^* \text{ if } o(q) = m + 1 \le n,$$

$$D_1 = \text{diag}(q^m, q^{m-1}, ..., q), \qquad \text{if } o(q) = 2m \le 2n,$$

$$D_2 = \text{diag}(a_1, a_2, ..., a_k) \in GL_k(\mathbb{C}) \qquad \text{for some } k \in \mathbb{N}.$$

We then can prove (b) as the case of type A_n.

(c). *Let D be as (6.11.1), then the following two conditions are equivalent.*

(i). $\mathbf{g}_{D,q} = \mathcal{N}_{D,q}$ *but $\mathcal{N}_{D,q}$ is not contained in $\mathbf{g}^+$.*

(ii). *$o(q) = 2n' - 1$ for some integer n' in $(n/2, n]$, and there exist some i in $[1, k]$ and m in $[0, m_i]$ such that $d_i q^m = q$, $2m_i - 2m + 2 > o(q)$.*

The proof is straight.

(d). (i). *If $o(q) > 2n$, , then we have $\mathbf{g}_{D,q} = \mathcal{N}_{D,q} \subset \mathbf{g}^+$.*

(ii). *Assume that $\mathbf{g}_{D,q} = \mathcal{N}_{D,q} \subset \mathbf{g}^+$, then for any $r \in \mathbb{C}^*_{W_0}$, there exists E in T such that $(D, q) \sim (E, r)$.*

Part (i) is trivial. The proof of part (ii) is similar to type A_n although a little more care is needed.

(e). *Let s be an element in T and q a primitive $2n$-th root of 1, then*

(i). *$\mathbf{g}_{s,q}$ is not equal to $\mathcal{N}_{s,q}$ if and only if s is conjugate to the following element*

$$D = \operatorname{diag}(1, q^n, q^{n-1}, \cdots, q^2, q, q^{-n}, q^{1-n}, \cdots, q^{-2}, q^{-1}).$$

(ii). *If $\mathbf{g}_{s,q} = \mathcal{N}_{s,q}$, then for any $r \in \mathbb{C}^*_{W_0}$ we can find some $t \in T$ such that $(s, q) \sim (t, r)$.*

Proof. (i) follows from (a). Since q is a primitive $2n$-th root of 1, $\mathbf{g}_{s,q} = \mathcal{N}_{s,q}$ implies that $\mathbf{g}_{s,q} \subset \dot{w}.\mathbf{g}^+$ for some w in W_0. Then (ii) follows from (d).

6.12. Type C_n. We consider

$$G = Sp_{2n}(\mathbb{C}) = \{g \in SL_{2n}(\mathbb{C}) \mid \tilde{g} \begin{pmatrix} 0 & I_n \\ -I_n & 0 \end{pmatrix} g = \begin{pmatrix} 0 & I_n \\ -I_n & 0 \end{pmatrix}\},$$

where $\tilde{g}$ is the transpose of g. We choose the maximal torus T to be the set of the diagonal matrices in G. Then

$$T = \{\operatorname{diag}(a_1, a_2, ..., a_n, a_1^{-1}, a_2^{-1}, ..., a_n^{-1}) \mid a_1, a_2, ..., a_n \in \mathbb{C}^*\}.$$

The normalizer $N_G(T)$ of T in G is generated by T and elements $P_{ij}P_{i+n,j+n}$, $P_i(-1)P_{i,i+n}$ ($1 \le i \ne j \le n$), where P_{ij} is the matrix obtained by exchanging the i-th row and the j-th row of the matrix I_{2n}, and $P_i(-1)$ is the matrix obtained by multiplying the i-th row of the matrix I_{2n} with -1. The Weyl group $W_0 = N_G(T)/T$ is isomorphic to the semi-direct product $(\mathbb{Z}/2\mathbb{Z})^n \ltimes \mathfrak{S}_n$.

Let $\alpha_{ij}, \beta_{ij}, \gamma_k \in X = \operatorname{Hom}(T, \mathbb{C}^*)$, $1 \le i < j \le n, 1 \le k \le n$, be defined as follows:

$$\alpha_{ij} : \operatorname{diag}(a_1, a_2, ..., a_n, a_1^{-1}, a_2^{-1}, ..., a_n^{-1}) \to a_i a_j^{-1},$$

70

$$\beta_{ij} : \mathrm{diag}\,(a_1, a_2, ..., a_n, a_1^{-1}, a_2^{-1}, ..., a_n^{-1}) \to a_i a_j,$$

$$\gamma_k : \mathrm{diag}\,(a_1, a_2, ..., a_n, a_1^{-1}, a_2^{-1}, ..., a_n^{-1}) \to a_k^2.$$

Let $R^+ = \{\alpha_{ij},\ \beta_{ij},\ \gamma_k \mid 1 \le i < j \le n,\ 1 \le k \le n\}$. Then the set of simple roots is $\Delta = \{\alpha_{i,i+1},\ \gamma_n \mid 1 \le i \le n - 1\}$.

Let (s, q) be an element in $T \times \mathbb{C}^*$ and assume that $q \ne 1$. Obviously through an element of the Weyl group W_0, s is conjugate to certain element $D \in T$ of the following form:

(6.12.1) $$D = \mathrm{diag}(D_1, D_2, ..., D_k, D_1^{-1}, D_2^{-1}, ..., D_k^{-1}),$$

where

$$D_i = \begin{pmatrix} d_i q^{m_i} I_{r_{i,i}} & 0 & \cdots & 0 & 0 \\ 0 & d_i q^{m_i-1} I_{r_{i,i-1}} & \cdots & 0 & 0 \\ \vdots & \vdots & \ddots & \vdots & \vdots \\ 0 & 0 & \cdots & d_i q I_{r_{i,1}} & 0 \\ 0 & 0 & \cdots & 0 & d_i I_{r_{i,0}} \end{pmatrix}, \qquad 1 \le i \le k,$$

all $r_{i,j}$ are positive integers, and all m_i are non-negative integers, $d_i \in \mathbb{C}^*$, moreover $\max\{m_1, m_2, ..., m_k\} < o(q)$, and

$$d_i q^m (d_j q^{m'})^{\pm 1} \ne 1, q, q^{-1}, \quad \text{for any } 1 \le i \ne j \le k,\ 0 \le m \le m_i,\ 0 \le m' \le m_j.$$

We have

(a). $\mathbf{g}_{s,q}$ is not equal to $\mathcal{N}_{s,q}$ if and only if at least one of the following conditions is satisfied.

(i). There is some integer i in $[1, k]$ such that $m_i + 1 = o(q)$.

(ii). $o(q)$ is even and there are some i in $[1, k]$ and m in $[0, m_i]$ such that $d_i q^m = q^{\frac{1}{2}}$ and $2m_i - 2m + 2 \ge o(q)$.

Proof. We may prove the result as the case of type A_n.

(b). If $\mathbf{g}_{s,q}$ is not equal to $\mathcal{N}_{s,q}$, then there exists a sequence $t_1, t_2, ..., t_k, ...$ in T_{reg} such that

$$\mathbf{g}_{t_k,q} \ne \mathcal{N}_{t_k,q}, \quad \text{and} \quad \lim_{k \to \infty} t_k = s.$$

Proof. Assume that $\mathbf{g}_{s,q}$ is not equal to $\mathcal{N}_{s,q}$. By (a) we see that s is conjugate to certain element $D = \mathrm{diag}(D_1, D_2, D_1^{-1}, D_2^{-1}) \in T$, such that

$$D_1 = \mathrm{diag}(dq^m, dq^{m-1}, ..., dq, d), \qquad \text{for some } d \in \mathbb{C}^* \text{ if } o(q) = m + 1 \le n,$$

$$D_1 = \mathrm{diag}(q^{\frac{2m-1}{2}}, q^{\frac{2m-3}{2}}, ..., q^{\frac{3}{2}}, q^{\frac{1}{2}}), \qquad \text{if } o(q) = 2m \le 2n,$$

$$D_2 = \mathrm{diag}(a_1, a_2, ..., a_k) \in GL_k(\mathbb{C}) \qquad \text{for some } k \in \mathbb{N}.$$

71

We then can prove (b) as the case of type A_n.

(c). *Let D be as (6.12.1), then the following two conditions are equivalent.*

(i). $\mathbf{g}_{D,q} = \mathcal{N}_{D,q}$ *but $\mathcal{N}_{D,q}$ is not contained in $\mathbf{g}^+$.*

(ii). $o(q) = 2n' - 1$ *for some integer n' in $(n/2, n]$, and there exist some i in $[1, k]$ and m in $[0, m_i]$ such that $d_i q^m = q^{\frac{1}{2}}$, $2m_i - 2m + 2 > o(q)$.*

The proof is straight.

(d). (i). *If $o(q) > 2n$, then we have $\mathbf{g}_{D,q} = \mathcal{N}_{D,q} \subset \mathbf{g}^+$.*

(ii). *Assume that $\mathbf{g}_{D,q} = \mathcal{N}_{D,q} \subset \mathbf{g}^+$, then for any $r \in \mathbb{C}^*_{W_0}$, there exists E in T such that $(D, q) \sim (E, r)$.*

Part (i) is trivial. Part (ii) is similar to case of type A_n.

(e). *Let $s \in T$. Assume that q is a primitive $2n$-th root of 1, then*

(i). $\mathbf{g}_{s,q}$ *is not equal to $\mathcal{N}_{s,q}$ if and only if s is conjugate to the following element*

$$D = \mathrm{diag}(q^{\frac{2n-1}{2}}, q^{\frac{2n-3}{2}}, \cdots, q^{\frac{3}{2}}, q^{\frac{1}{2}}, q^{\frac{1-2n}{2}}, q^{\frac{3-2n}{2}}, \cdots, q^{-\frac{3}{2}}, q^{-\frac{1}{2}}).$$

(ii). *If $\mathbf{g}_{s,q} = \mathcal{N}_{s,q}$, then for any $r \in \mathbb{C}^*_{W_0}$, we can find some $t \in T$ such that $(s, q) \sim (t, r)$.*

Proof. (i) follows from (a). Since q is a primitive $2n$-th root of 1, $\mathbf{g}_{s,q} = \mathcal{N}_{s,q}$ implies that $\mathbf{g}_{s,q} \subset \dot{w}.\mathbf{g}^+$ for some w in W_0. Then (ii) follows from (d).

6.13. Type D_n. We consider

$$G = SO_{2n}(\mathbb{C}) = \{g \in SL_{2n}(\mathbb{C}) \mid \tilde{g} \begin{pmatrix} 0 & I_n \\ I_n & 0 \end{pmatrix} g = \begin{pmatrix} 0 & I_n \\ I_n & 0 \end{pmatrix}\},$$

where $\tilde{g}$ is the transpose of g. We choose the maximal torus T to be the set of the diagonal matrices in G. Then

$$T = \{\mathrm{diag}(a_1, a_2, ..., a_n, a_1^{-1}, a_2^{-1}, ..., a_n^{-1}) \mid a_1, a_2, ..., a_n \in \mathbb{C}^*\}.$$

The normalizer $N_G(T)$ of T in G is generated by T and elements $P_{ij}P_{i+n,j+n}$, $P_{i,i+n}P_{j,j+n}$ $(1 \le i \ne j \le n)$, where P_{ij} is the matrix obtained by exchanging the i-th row and the j-th row of the matrix I_{2n}. The Weyl group $W_0 = N_G(T)/T$ is isomorphic to the semi-direct product $(\mathbb{Z}/2\mathbb{Z})^{n-1} \ltimes \mathfrak{S}_n$.

Let $\alpha_{ij}, \beta_{ij} \in X = \mathrm{Hom}(T, \mathbb{C}^*)$, $1 \le i < j \le n$, be defined as follows.

$$\alpha_{ij} : \mathrm{diag}\,(a_1, a_2, ..., a_n, a_1^{-1}, a_2^{-1}, ..., a_n^{-1}) \to a_i a_j^{-1},$$

$$\beta_{ij} : \mathrm{diag}\,(a_1, a_2, ..., a_n, a_1^{-1}, a_2^{-1}, ..., a_n^{-1}) \to a_i a_j.$$

Let $R^+ = \{\alpha_{ij}, \beta_{ij} \mid 1 \leq i < j \leq n\}$. Then the set of simple roots is $\Delta = \{\alpha_{i,i+1}, \beta_{n-1,n} \mid 1 \leq i \leq n-1\}$.

Let (s,q) be an element in $T \times \mathbb{C}^*$ and assume that $q \neq 1$. Obviously through an element of the Weyl group W_0, s is conjugate to certain element $D \in T$ of the following form:

$$(6.13.1) \qquad D = \mathrm{diag}(D_1, D_2, ..., D_k, D_1^{-1}, D_2^{-1}, ..., D_k^{-1}),$$

where

$$D_i = \begin{pmatrix} d_i q^{m_i} I_{r_{i,i}} & 0 & \cdots & 0 & 0 \\ 0 & d_i q^{m_i-1} I_{r_{i,i-1}} & \cdots & 0 & 0 \\ \vdots & \vdots & \ddots & \vdots & \vdots \\ 0 & 0 & \cdots & d_i q I_{r_{i,1}} & 0 \\ 0 & 0 & \cdots & 0 & d_i I_{r_{i,0}} \end{pmatrix}, \qquad 1 \leq i \leq k,$$

all $r_{i,j}$ are positive integers, and all m_i are non-negative integers, $d_i \in \mathbb{C}^*$, moreover $\max\{m_1, m_2, ..., m_k\} < o(q)$ and

$$(6.13.2)$$
$$d_i q^m (d_j q^{m'})^{\pm 1} \neq 1, q, q^{-1}, \qquad \text{for any } 1 \leq i \neq j \leq k, 0 \leq m \leq m_i, 0 \leq m' \leq m_j;$$

or

$$(6.13.3)$$
$$d_i q^m (d_j q^{m'})^{\pm 1} \neq 1, q, q^{-1}, \qquad \text{if } i \neq j, \text{ and } i \notin \{k-1, k\} \text{ or } j \notin \{k-1, k\},$$

$0 \leq m \leq m_i$, $0 \leq m' \leq m_j$, and $D_k = (d_k)$, $d_{k-1} q^m d_k^{-1} \neq 1, q$ for $m = 0, 1, ..., m_{k-1}$, and $d_{k-1} q^l d_k = q$ for some integer l in $[1, m_{k-1}]$. We also require that m_{k-1} is as big as possible.

(a). $\mathbf{g}_{s,q}$ is not equal to $\mathcal{N}_{s,q}$ if and only if at least one of the following conditions is satisfied.

(i).. There is some integer i in $[1, k]$ such that $m_i + 1 = o(q)$.

(ii). $o(q)$ is even and there are some i in $[1, k]$ and m in $[0, m_i]$ such that $d_i q^m = 1$ and $2m_i - 2m + 2 \geq o(q)$.

Proof. We may prove the result as the case of type A_n. The results in [C1] and 6.9(b) are helpful in the proof.

(b). If $\mathbf{g}_{s,q}$ is not equal to $\mathcal{N}_{s,q}$, then there exists a sequence $t_1, t_2, ..., t_k, ...$ in T_{reg} such that

$$\mathbf{g}_{t_k,q} \neq \mathcal{N}_{t_k,q}, \quad \text{and} \quad \lim_{k \to \infty} t_k = s.$$

Proof. Assume that $\mathbf{g}_{s,q}$ is not equal to $\mathcal{N}_{s,q}$. By (a) we see that s is conjugate to certain element $D = \mathrm{diag}(D_1, D_2, D_1^{-1}, D_2^{-1}) \in T$, such that

$$D_1 = \mathrm{diag}(dq^m, dq^{m-1}, ..., dq, d), \qquad \text{for some } d \in \mathbb{C}^* \text{ if } o(q) = m+1 \leq n,$$

73

$$D_1 = \operatorname{diag}(q^m, q^{m-1}, ..., q, 1), \quad \text{if } o(q) = 2m - 2 < 2n,$$

$$D_2 = \operatorname{diag}(a_1, a_2, ..., a_k) \in GL_k(\mathbb{C}) \quad \text{for some } k \in \mathbb{N}.$$

We then can prove (b) as the case of type A_n.

(c). *Let D be as (6.13.1), then the following two conditions are equivalent.*

(i). $\mathbf{g}_{D,q} = \mathcal{N}_{D,q}$ *but* $\mathcal{N}_{D,q}$ *is not contained in* $\mathbf{g}^+$.

(ii). $o(q) = 2n' - 1$ *for some n' in $(n/2, n-1]$, and there exist some i in $[1, k]$ and m in $[0, m_i]$ such that $d_i q^m = 1$, $2m_i - 2m + 2 > o(q)$.*

The proof is straight.

(d). (i). *If $o(q) > 2n - 2$, then we have $\mathbf{g}_{D,q} = \mathcal{N}_{D,q} \subset \mathbf{g}^+$.*

(ii). *Assume that $\mathbf{g}_{D,q} = \mathcal{N}_{D,q} \subset \mathbf{g}^+$, then for any $r \in \mathbb{C}_{W_0}^*$, there exists E in T such that $(D, q) \sim (E, r)$.*

Part (i) is trivial. Part (ii) is similar to case of type A_n but more tedious.

(e). *Let $s \in T$. Assume that q is a primitive $(2n - 2)$-th root of 1, then*

(i). $\mathbf{g}_{s,q}$ *is not equal to $\mathcal{N}_{s,q}$ if and only if s is conjugate to the following element*

$$\operatorname{diag}(q^{n-1}, q^{n-2}, ..., q, 1, q^{1-n}, q^{2-n}, ..., q^{-1}, 1).$$

(ii). *If $\mathbf{g}_{s,q} = \mathcal{N}_{s,q}$, then for any $r \in \mathbb{C}_{W_0}^*$, we can find t in T such that $(s, q) \sim (t, r)$.*

Proof. (i) follows from (a). Since q is a primitive $(2n - 2)$-th root of 1, $\mathbf{g}_{s,q} = \mathcal{N}_{s,q}$ implies that $\mathbf{g}_{s,q} \subset \dot{w}.\mathbf{g}^+$ for some w in W_0. Then (ii) follows from (d).

6.14. Exceptioal types.

Let G be a simple algebraic group of adjoint type, then

(6.14.1) $T \simeq \operatorname{Hom}(P, \mathbb{C}^*)$, where T is a maximal torus in G and $P = \operatorname{Hom}(T, \mathbb{C}^*)$ is the character group of T. Note that P is also the root lattice.

There are no simple realizations for algebraic groups of exceptional types, so for us the property (6.14.1) is important. To use it we need explicit structures of the root systems of exceptional types. We adopt the approach in [OV]. For type F_4, the approach is the same as in [B].

Type E_6: Let $\varepsilon_1, \varepsilon_2, ..., \varepsilon_6$ be vectors in $\mathbb{R}^6$ satisfying $\sum \varepsilon_i = 0$ and

$$(\varepsilon_i, \varepsilon_i) = 5/6, \quad (\varepsilon_i, \varepsilon_j) = -1/6 \quad \text{for } i \neq j.$$

Let $\varepsilon \in \mathbb{R}^6$ be such that $(\varepsilon, \varepsilon_i) = 0$ for all i and $(\varepsilon, \varepsilon) = 1/2$.

Type E_7, E_8, G_2: Let $\varepsilon_1, \varepsilon_2, ..., \varepsilon_{n+1}$ ($n = $ rank) be vectors in $\mathbb{R}^{n+1}$ satisfying $\sum \varepsilon_i = 0$ and

$$(\varepsilon_i, \varepsilon_i) = n/(n+1), \quad (\varepsilon_i, \varepsilon_j) = -1/(n+1) \quad \text{for } i \neq j.$$

74

Type F_4: Let $\varepsilon_1, \varepsilon_2, \varepsilon_3, \varepsilon_4$ be an orthonormal basis of $\mathbb{R}^4$.

Then we have

(a) Type E_6. The roots are: $\varepsilon_i - \varepsilon_j$, $\pm 2\varepsilon$, $\varepsilon_i + \varepsilon_j + \varepsilon_k \pm \varepsilon$. We choose $\varepsilon_i - \varepsilon_{i+1}$ ($i <$ 6), $\varepsilon_4 + \varepsilon_5 + \varepsilon_6 + \varepsilon$ as the set of simple roots.

(b) Type E_7. The roots are $\varepsilon_i - \varepsilon_j$, $\varepsilon_i + \varepsilon_j + \varepsilon_k + \varepsilon_l$. We choose $\varepsilon_i - \varepsilon_{i+1}$ ($i <$ 7), $\varepsilon_5 + \varepsilon_6 + \varepsilon_7 + \varepsilon_8$ as the set of simple roots.

(c) Type E_8. The roots are $\varepsilon_i - \varepsilon_j$, $\pm(\varepsilon_i + \varepsilon_j + \varepsilon_k)$. We choose $\varepsilon_i - \varepsilon_{i+1}$ ($i <$ 8), $\varepsilon_6 + \varepsilon_7 + \varepsilon_8$ as the set of simple roots.

(d) Type F_4. The roots are $\pm \varepsilon_i \pm \varepsilon_j$, $\pm \varepsilon_i$, $(\pm \varepsilon_1 \pm \varepsilon_2 \pm \varepsilon_3 \pm \varepsilon_4)/2$. We choose $(\varepsilon_1 - \varepsilon_2 - \varepsilon - 3 - \varepsilon_4)/2$, ε_4, $\varepsilon_3 - \varepsilon_4$, $\varepsilon_2 - \varepsilon_3$ as the set of simple roots.

(e) Type G_2. The roots are: $\varepsilon_i - \varepsilon_j$, $\pm \varepsilon_i$. We choose ε_2, $\varepsilon_3 - \varepsilon_2$ as the set of simple roots.

It is convenient to present the formulas for f_{W_0}. We have

Type	f_{W_0}
E_8	$\dfrac{(\mathbf{q}^{30} - 1)(\mathbf{q}^{24} - 1)(\mathbf{q}^{20} - 1)(\mathbf{q}^{18} - 1)(\mathbf{q}^{14} - 1)(\mathbf{q}^{12} - 1)(\mathbf{q}^8 - 1)(\mathbf{q}^2 - 1)}{(\mathbf{q} - 1)^8}$
E_7	$\dfrac{(\mathbf{q}^{18} - 1)(\mathbf{q}^{14} - 1)(\mathbf{q}^{12} - 1)(\mathbf{q}^{10} - 1)(\mathbf{q}^8 - 1)(\mathbf{q}^6 - 1)(\mathbf{q}^2 - 1)}{(\mathbf{q} - 1)^7}$
E_6	$\dfrac{(\mathbf{q}^{12} - 1)(\mathbf{q}^9 - 1)(\mathbf{q}^8 - 1)(\mathbf{q}^6 - 1)(\mathbf{q}^5 - 1)(\mathbf{q}^2 - 1)}{(\mathbf{q} - 1)^6}$
F_4	$\dfrac{(\mathbf{q}^{12} - 1)(\mathbf{q}^8 - 1)(\mathbf{q}^6 - 1)(\mathbf{q}^2 - 1)}{(\mathbf{q} - 1)^4}$
G_2	$\dfrac{(\mathbf{q}^6 - 1)(\mathbf{q}^2 - 1)}{(\mathbf{q} - 1)^2}$

Now using (6.14.1), (a-e) and 6.9(b) we can prove the following results through lengthy case by case analysis.

Assume that G is a simple algebraic group of exceptional type and of adjoint type. We have

(f). If $\mathbf{g}_{s,q}$ is not equal to $\mathcal{N}_{s,q}$, then

(i). There exists a sequence $t_1, t_2, ..., t_k, ...$ in T_{reg} such that

$$\mathbf{g}_{t_k, q} \neq \mathcal{N}_{t_k, q}, \quad \text{and} \quad \lim_{k \to \infty} t_k = s.$$

or

(ii). *There exists a sequence $t_1, t_2, ..., t_k, ...$ in T_{reg} such that $\lim_{k \to \infty} t_k = s$ and for any $w \in W_0$ we have*

$$\lim_{k \to \infty} \prod_{\alpha \in R^+} \frac{1 - qw(\alpha)(t_k)}{1 - w(\alpha)(t_k)} = 0.$$

Proof. We shall freely use the results on the conjugacy classes in Weyl groups in [C1]. We number the simple roots according to the Coxeter graphs in 1.3. By means of the adjoint representations and using 6.9(b) and the results in [C1] we see that $g_{s,q}$ is not equal to $\mathcal{N}_{s,q}$ $(s \in T)$ if and only if s is conjugate to an element t in T which satisfys one of the following conditions.

Type E_8:

$$\alpha_8(t) = -1, \qquad q = -1$$
$$\alpha_8(t) = \alpha_7(t) = q, \qquad o(q) = 3$$
$$\alpha_8(t) = \alpha_7(t) = \alpha_6(t) = q, \qquad o(q) = 4$$
$$\alpha_i(t) = q, \quad i = 5, 6, 7, 8 \qquad o(q) = 5$$
$$\alpha_i(t) = q, \quad 4 \leq i \leq 8 \qquad o(q) = 6$$
$$\alpha_i(t) = q, \quad 2 \leq i \leq 5 \qquad o(q) = 6$$
$$\alpha_i(t) = q, \quad 3 \leq i \leq 8 \qquad o(q) = 7$$
$$\alpha_i(t) = q, \quad 1 \leq i \leq 5 \qquad o(q) = 8$$
$$\alpha_i(t) = q, \quad 3 \leq i \leq 8 \text{ or } i = 1 \qquad o(q) = 8$$
$$\alpha_i(t) = q, \quad 3 \leq i \leq 8 \text{ or } i = 1 \qquad o(q) = 9$$
$$\text{and } \alpha_2(t) = q^3$$
$$\alpha_i(t) = q, \quad 1 \leq i \leq 6 \qquad o(q) = 9$$
$$\alpha_i(t) = q, \quad 2 \leq i \leq 7 \qquad o(q) = 10$$
$$\alpha_i(t) = q, \quad 1 \leq i \leq 6 \qquad o(q) = 12$$
$$\alpha_i(t) = q, \quad 2 \leq i \leq 8 \qquad o(q) = 12$$
$$\alpha_i(t) = q, \quad 1 \leq i \leq 7 \qquad o(q) = 14$$
$$\alpha_i(t) = q, \quad 2 \leq i \leq 8 \qquad o(q) = 14$$
$$\text{and } \alpha_1(t) = \pm 1 \qquad o(q) = 14$$
$$\alpha_i(t) = q, \quad 1 \leq i \leq 7 \qquad o(q) = 18$$
$$\alpha_i(t) = q, \quad 1 \leq i \leq 8 \qquad o(q) = 20$$
$$\alpha_i(t) = q, \quad 1 \leq i \leq 8 \qquad o(q) = 24$$
$$\alpha_i(t) = q, \quad 1 \leq i \leq 8 \qquad o(q) = 30$$

Type E_7:

$$\alpha_7(t) = -1, \qquad q = -1$$
$$\alpha_7(t) = \alpha_6(t) = q, \qquad o(q) = 3$$
$$\alpha_7(t) = \alpha_6(t) = \alpha_5(t) = q, \qquad o(q) = 4$$
$$\alpha_i(t) = q, \quad i = 4, 5, 6, 7 \qquad o(q) = 5$$

$$\alpha_i(t) = q, \quad 3 \le i \le 7 \qquad\qquad o(q) = 6$$
$$\alpha_i(t) = q, \quad 2 \le i \le 5 \qquad\qquad o(q) = 6$$
$$\alpha_i(t) = q, \quad 3 \le i \le 7 \text{ or } i = 1 \qquad o(q) = 7$$
$$\alpha_i(t) = q, \quad 3 \le i \le 7 \text{ or } i = 1 \qquad o(q) = 8$$
$$\text{and } \alpha_2(t) = 1$$
$$\alpha_i(t) = q, \quad 1 \le i \le 5 \qquad\qquad o(q) = 8$$
$$\alpha_i(t) = q, \quad 1 \le i \le 6 \qquad\qquad o(q) = 9$$
$$\alpha_i(t) = q, \quad 2 \le i \le 7 \qquad\qquad o(q) = 10$$
$$\alpha_i(t) = q, \quad 1 \le i \le 6 \qquad\qquad o(q) = 12$$
$$\alpha_i(t) = q, \quad 1 \le i \le 7 \qquad\qquad o(q) = 14$$
$$\alpha_i(t) = q, \quad 1 \le i \le 7 \qquad\qquad o(q) = 18$$

Type E_6:

$$\alpha_6(t) = -1, \qquad\qquad q = -1$$
$$\alpha_6(t) = \alpha_5(t) = q, \qquad\qquad o(q) = 3$$
$$\alpha_6(t) = \alpha_5(t) = \alpha_4(t) = q, \qquad o(q) = 4$$
$$\alpha_i(t) = q, \quad i = 3,4,5,6 \qquad\qquad o(q) = 5$$
$$\alpha_i(t) = q, \quad 3 \le i \le 6 \text{ or } i = 1 \qquad o(q) = 6$$
$$\alpha_i(t) = q, \quad 2 \le i \le 5 \qquad\qquad o(q) = 6$$
$$\alpha_i(t) = q, \quad 1 \le i \le 5 \qquad\qquad o(q) = 8$$
$$\alpha_i(t) = q, \quad 1 \le i \le 6 \qquad\qquad o(q) = 9$$
$$\alpha_i(t) = q, \quad 1 \le i \le 6 \qquad\qquad o(q) = 12$$

Type F_4:

$$\alpha_1(t) = -1 \text{ or } \alpha_4 = -1, \qquad\qquad q = -1$$
$$\alpha_1(t) = \alpha_2(t) = q \text{ or } \alpha_3(t) = \alpha_4(t) = q, \qquad o(q) = 3$$
$$\alpha_2(t) = \alpha_3(t) = q, \qquad\qquad o(q) = 4$$
$$\alpha_1(t) = \alpha_2(t) = q, \ \alpha_3(t)\alpha_4(t) = \pm 1, \qquad o(q) = 4$$
$$\alpha_i(t) = q, \quad i = 3,4,5,6 \qquad\qquad o(q) = 5$$
$$\alpha_i(t) = q, \quad 1 \le i \le 3 \text{ or } i = 1 \qquad o(q) = 6$$
$$\alpha_i(t) = q, \quad 2 \le i \le 4 \qquad\qquad o(q) = 6$$
$$\alpha_i(t) = q, \quad 1 \le i \le 4 \qquad\qquad o(q) = 8$$
$$\alpha_i(t) = q, \quad 1 \le i \le 4 \qquad\qquad o(q) = 12$$

Type G_2:

$$\alpha_1(t) = -1 \text{ or } \alpha_2(t) = -1, \qquad\qquad q = -1$$
$$\alpha_1(t) = \alpha_2(q) = q, \qquad\qquad o(q) = 8$$
$$\alpha_1(t) = \alpha_2(t) = q, \qquad\qquad o(q) = 12$$

Thus it is sufficient to prove (f) for those $t \in T$ satisfying the conditions in the tables. We use type G_2 as an example to prove it. We identify an element $r \in T$ with the pair $(\alpha_1(r), \alpha_2(r))$.

If $q = -1$ and $\alpha_1(t) = -1$, we choose a sequence $a_1, a_2, ..., a_k, ...$ of positive real numbers such that $\lim_{k \to \infty} a_k = 1$ and such that all $t_k = (-1, a_k \alpha_2(t))$ are contained in T_{reg} (this is possible by a simple calculation). Then $\mathbf{g}_{t_k, -1}$ is not equal to $\mathcal{N}_{t_k, -1}$ and $\lim_{k \to \infty} t_k = t$. Similarly we deal with the case $\alpha_2(t) = -1$.

If $o(q) = 3$, then $\alpha_1(t) = \alpha_2(t) = q$. We choose a sequence $a_1, a_2, ..., a_k, ...$ of positive real numbers in the open interval $(0,1)$ such that $\lim_{k \to \infty} a_k = 1$. Then every $t_k = (a_k q, a_k q)$ is contained in T_{reg} and $\lim_{k \to \infty} t_k = t$. For an arbitrary $w \in W_0$, one may check that

$$\lim_{k \to \infty} \prod_{\alpha \in R^+} \frac{1 - q w(\alpha)(t_k)}{1 - w(\alpha)(t_k)} = 0.$$

If $o(q) = 6$, then $\alpha_1(t) = \alpha_2(t) = q$. The element t is regular and for any $w \in W_0$ we have

$$\prod_{\alpha \in R^+} \frac{1 - q w(\alpha)(t)}{1 - w(\alpha)(t)} = 0.$$

We can deal with other types in a similar way. Thus we complete the proof.

(g). *Let* $(s, q) \in T \times \mathbb{C}^*$.

(i). *If* $o(q) > e_n + 1$, , *then we have* $\mathbf{g}_{w(s), q} = \mathcal{N}_{w(s), q} \subset \mathbf{g}^+$ *for some* $w \in W_0$.

(ii). *Assume that* $\mathbf{g}_{s, q} = \mathcal{N}_{s, q} \subset \mathbf{g}^+$, *then for any* $r \in \mathbb{C}^*_{W_0}$, *there exists* t *in* T *such that* $R_{s, q} = R_{t, r}$ *and* $R_{s, 1} = R_{t, 1}$.

Proof. (i). It is equivalent to prove that $w^{-1}(R_{s, q}) \subset R^+$ for some $w \in W_0$. Let $R^+_{s, q} = R_{s, q} \cap R^+$ and $R^-_{s, q} = R_{s, q} \cap R^-$. If $R_{s, q} = R^+_{s, q}$, nothing need to argue since we can choose $w = e$, the neutral element in W_0. Now assume that $R_{s, q}$ is not equal to $R^+_{s, q}$. Note that $o(q) > e_n + 1$, we see that the subgroup of the root lattice P generated by $R^+_{s, q}$ doesnot contain any element of $R^-_{s, q}$. Choose $\beta \in R^-_{s, q}$ and let w_1 be the reflection corresponding to β. Then $w_1(\beta) \in R^+$ and $w_1(R^+_{s, q}) \subset R^+$. Thus $|w_1(R^+_{s, q})| = |R^+_{w_1(s), q}| > |R^+_{s, q}|$. We now can use induction on $|R^+_{s, q}|$ since $|R_{w_1(s), q}| = |R_{s, q}|$.

(ii). We can prove the assertion case by case. We omit the tedious proof.

(h). *Let* s *be an element in* T *and* q *a primitive* $(e_n + 1)$-*th root of* 1, *then*

(i). $\mathbf{g}_{s, q}$ *is not equal to* $\mathcal{N}_{s, q}$ *if and only if* s *is conjugate to the an element* t *such that* $\alpha(t) = q$ *for any simple root* α.

(ii). *If* $\mathbf{g}_{s, q} = \mathcal{N}_{s, q}$, *then for any* $r \in \mathbb{C}^*_{W_0}$, *we can find* t *in* T *such that* $R_{s, q} = R_{t, r}$ *and* $R_{s, 1} = R_{t, 1}$.

Proof. (i). It follows from the proof of part (i) of (f). Using the proof of (f) and the proof of part (i) of (g) we see that if $\mathbf{g}_{s, q} = \mathcal{N}_{s, q}$, then we can find $w \in W_0$ such that $\mathbf{g}_{w(s), q} = \mathcal{N}_{w(s), q} \subset \mathbf{g}^+$. Thus (ii) follows from (g).

(j). Let $(s,q) \in T \times \mathbb{C}^*$ be such that $\mathbf{g}_{s,q} = \mathcal{N}_{s,q}$, then we can find (t,r) in $T \times \mathbb{C}^*$ such that $f_{W_0}(r) \neq 0$ and $R_{s,q} = R_{t,r}$ and $R_{s,1} = R_{t,1}$.

Proof. Use the table and case by case analysis. We omit the details.

6.15. Now we can see that 6.5 and 6.8 follow from the results in 6.10-6.14, and 6.1 (j-k).

It would be interesting to find a necessary and sufficient condition for the natural isomorphism between $\mathbf{H}_{s,q}$ and $\mathbf{H}_{t,r}$.

6.16. Conjecture. Assume that G has a simply connected derived group, then $\mathbf{H}_{s,q}$ is isomorphic to $\mathbf{H}_{t,r}$ provided that $\mathcal{N}_{s,q} = \mathcal{N}_{t,r}$.

7. The Lowest Two-Sided Cell

Notations are as in chapter 5. It is known that $\mathbf{K}_{G \times \mathbb{C}^*}(\mathcal{B} \times \mathcal{B})$ may be regarded as a two-sided ideal of the algebra $\mathbf{K}_{G \times \mathbb{C}^*}(Z)$. In this chapter we will give an explicit description for the ideal. Another purpose is to show that 5.6(d) is not true when $f_{w_0}(q) = 0$. For simplicity we assume that G is a simply connected, simple algebraic group over $\mathbb{C}$. All these are done by using the knowledge concerned with the lowest two-sided cell

$$c_0 = \{w \in W \mid a(w) = l(w_0)\}.$$

The two-sided cell c_0 corresponds to the nilpotent G-orbit $\{0\}$ under Lusztig's bijection between the set $\mathrm{Cell}(W)$ of two-sided cells of W and the set of nilpotent G-orbits in $\mathbf{g}$.

7.1. The ideal $\mathbf{K}_{G \times \mathbb{C}^*}(\mathcal{B} \times \mathcal{B})$ of $\mathbf{K}_{G \times \mathbb{C}^*}(Z)$. For an arbitrary nilpotent G-orbit $\mathcal{C}$, let

$$Z_{\bar{\mathcal{C}}} = \{(N, \mathbf{b}, \mathbf{b}') \in Z \mid N \in \bar{\mathcal{C}}\},$$

where $\bar{\mathcal{C}}$ is the closure of $\mathcal{C}$. The variety $Z_{\bar{\mathcal{C}}}$ is $G \times \mathbb{C}^*$-stable. It is known that the inclusion $Z_{\bar{\mathcal{C}}} \hookrightarrow Z$ induces an injection

$$\mathbf{K}_{G \times \mathbb{C}^*}(Z_{\bar{\mathcal{C}}}) \hookrightarrow \mathbf{K}_{G \times \mathbb{C}^*}(Z)$$

and the image is a two-sided ideal of the convolution algebra $\mathbf{K}_{G \times \mathbb{C}^*}(Z)$ (see [KL4]). We denote the image again by $\mathbf{K}_{G \times \mathbb{C}^*}(Z_{\bar{\mathcal{C}}})$. It is conjectured that the ideal is closely related to the two-sided cell corresponding to the nilpotent G-orbit $\mathcal{C}$ (see [Du, p.32; G4]). We shall give an explicit description of the ideal when $\mathcal{C}$ is the class $\{0\}$ (see Theorem 7.4).

We shall identify $\mathbf{K}_{G \times \mathbb{C}^*}(\mathcal{B})$ with $\mathbf{A}[X]$. Let $\mathbf{A}[X]^{W_0}$ be the set of W_0-invariant elements in $\mathbf{A}[X]$. It is known that $\mathbf{A}[X]^{W_0} = \mathbf{R}_{G \times \mathbb{C}^*} = \mathbf{A} \otimes_{\mathbb{C}} \mathbf{R}_G$. We have (see [KL4])

(a) The external tensor product in K-theory defines an isomorphism

$$\boxtimes : \mathbf{K}_{G \times \mathbb{C}^*}(\mathcal{B} \times \mathcal{B}) \simeq \mathbf{A}[X] \underset{\mathbf{A}[X]^{W_0}}{\otimes} \mathbf{A}[X]$$

as $\mathbf{A}[X]^{W_0}$-modules.

We shall identify $\mathbf{K}_{G \times \mathbb{C}^*}(\mathcal{B} \times \mathcal{B})$ with $\mathbf{A}[X] \underset{\mathbf{A}[X]^{W_0}}{\otimes} \mathbf{A}[X]$, and regard them as a two-sided ideal of the algebra $\mathbf{K}_{G \times \mathbb{C}^*}(Z)$.

(b) There exists a unique left $\dot{\mathbf{H}}$-module structure (denoted $h \circ \xi$) on $\mathbf{A}[X]$ such that

$$T_s \circ x = \frac{s(x) - x \alpha_s}{\alpha_s - 1} + \mathbf{q} \frac{x \alpha_s - s(x) \alpha_s^{-1}}{\alpha_s - 1}, \qquad (s \in S_0, \ x \in X),$$

80

$$\theta_{x_1} \circ x = x_1 x, \qquad (x_1, x \in X).$$

The action $\circ$ is a $\mathbf{q}$-analogue of the usual 'dot' action of W on $\mathbf{A}[X]$.

(c) In $\mathbf{K}_{G \times \mathbb{C}^{\bullet}}(Z) = \dot{\mathbf{H}}$ we have

$$h(x \boxtimes y) = h \circ x \boxtimes y, \qquad (x \boxtimes y)h = x \boxtimes h \circ y, \qquad h \in \mathbf{K}_{G \times \mathbb{C}^{\bullet}}(Z), \; x, y \in X.$$

7.2. Lemma. *There is a unique left $\dot{\mathbf{H}}$-module structure (denoted $h * \xi$) on $\mathbf{A}[X]$ extending the obvious $\mathbf{A}$ action and such that*

$$T_s * x = \frac{\alpha_s s(x) - x\alpha_s}{\alpha_s - 1} + \mathbf{q}\frac{x\alpha_s - s(x)}{\alpha_s - 1}, \qquad (s \in S_0, \; x \in X),$$

$$\theta_{x_1} * x = x_1 x, \qquad (x_1, x \in X).$$

Proof. We use Kato's trick to prove it. Let $\mathbf{I}$ be the left ideal of $\dot{\mathbf{H}}$ generated by $\sum_{w \in W_0} T_w$. Then the $\mathbf{A}$-linear map $\mathbf{A}[X] \to \mathbf{I}$ defined by $x \to \theta_x \sum_{w \in W_0} T_w$ is an isomorphism by 2.2(b). One checks easily that under this isomprphism the $\dot{\mathbf{H}}$-action $*$ becomes the left $\dot{\mathbf{H}}$-multiplication on $\mathbf{I}$. The lemma is proved.

It is easy to see that the action $*$ is a $\mathbf{q}$-analogue of the usual action of W on $\mathbf{A}[X]$. The relation between the actions $\circ$ and $*$ is simple. Let δ be the product of all fundamental weights. (Then δ^2 is the product of all positive roots in R. Note that here the operation in $X = \mathrm{Hom}(T, \mathbb{C}^*)$ is written multiplicatively.)

7.3. Lemma. *For $h \in \dot{\mathbf{H}}$ and $x \in X$, we have*

$$h \circ x = (\theta_\delta^{-1} h \theta_\delta) * x.$$

The proof is straight.

7.4. Theorem. *Let $\dot{\mathbf{H}}_{c_0}$ be the two-sided ideal of $\dot{\mathbf{H}}$ generated by $\sum_{w \in W_0} T_w$. It has natural $\dot{\mathbf{H}}$-bimodule structure through left and right multiplications. The map*

$$\mathbf{K}_{G \times \mathbb{C}^{\bullet}}(\mathcal{B} \times \mathcal{B}) \simeq \mathbf{A}[X] \otimes_{\mathbf{A}[X]^{W_0}} \mathbf{A}[X] \to \dot{\mathbf{H}}_{c_0}$$

defined by $x \boxtimes y \to \theta_\delta \theta_x \sum_{w \in W_0} T_w \theta_y \theta_\delta$ is an $\dot{\mathbf{H}}$-bimodule isomorphism.

Proof. According to 7.1(c) and (7.2-3), the map is a homomorphism of $\dot{\mathbf{H}}$-bimodule. Using 2.6(i) we see that the map is injective and surjective.

7.5. Now we shall classify the simple $\mathbf{H}_q$-modules attached to c_0. Recall the concept of attached two-sided cell in 5.9. For each semisimple element s in G, it is known (see [X3]) that at most one simple $\mathbf{H}_q$-module (up to isomorphisms) attached to c_0 such that U_x, $x \in X^+$ acts on it by scalar $tr(s, V(x))$. That is $|Y_{s,q,c_0}| \leq 1$.

We shall give a necessary and sufficient condition for $|Y_{s,q,c_0}| = 0$. We need some preparations.

Let $C = \sum_{w \in W_0} T_w$. For each $x \in X$, we write

$$W_x = \{w \in W_0 \mid w(x) = x\},$$

and

$$f_{W_x} = \sum_{w \in W_x} q^{l(w)}.$$

We shall need a result of Kato [Ka2] (see also [Gu]).

(a) If $x \in X^+$, then

$$C\theta_x C = f_{W_x} \sum_{w \in W_0} w(\theta_x \prod_{\alpha \in R^+} \frac{1 - q\theta_\alpha}{1 - \theta_\alpha})C.$$

We shall write $M_{s,q}$ instead of the standard module $M_{s,0,q,1}$. It is known that (see [KL4])

$$(7.5.1) \qquad M_{s,q} \simeq \mathbb{C}_{s,q} \otimes_{\mathbf{R}_{G \times \mathbf{C}^\bullet}} \mathbf{K}_{G \times \mathbf{C}^\bullet}(\mathcal{B}),$$

where $\dot{\mathbf{H}}$ acts on $\mathbf{K}_{G \times \mathbf{C}^\bullet}(\mathcal{B}) = \mathbf{A}[X]$ by $\circ$ (see 7.1(b)).

7.6. Lemma. *Let* $\mathbf{I}$ *be the left ideal of* $\mathbf{H}_q$ *generated by* $C = \sum_{w \in W_0} T_w$, *and let* $\mathbf{I}_s$ *be the left ideal of* $\mathbf{H}_q$ *generated by* $(U_x - tr(s, V(x))C$, $x \in X^+$, *then the quotient* $\mathbf{I}/\mathbf{I}_s$ *is just the standard module* $M_{s,q}$.

Proof. Using (7.3) and (7.5.1) we see that $M_{s,q} \simeq \mathbb{C}_{s,q} \otimes_{\mathbf{R}_{G \times \mathbf{C}^\bullet}} \mathbf{K}_{G \times \mathbf{C}^\bullet}(\mathcal{B})$, where $\dot{\mathbf{H}}$ acts on $\mathbf{K}_{G \times \mathbf{C}^\bullet}(\mathcal{B}) = \mathbf{A}[X]$ by $*$ (see (7.2)). By the definition of $*$ we see that

$$\mathbf{I}/\mathbf{I}_s \simeq \mathbb{C}_{s,q} \otimes_{\mathbf{R}_{G \times \mathbf{C}^\bullet}} \mathbf{K}_{G \times \mathbf{C}^\bullet}(\mathcal{B}).$$

The lemma is proved.

7.7. It is proved in [X3] that

(a) The set Y_{s,q,c_0} is empty if and only if $CM_{s,q} = 0$.

According to (7.6) we know that $CM_{s,q} = 0$ is equivalent to

(b) $C\theta_x C \in \mathbf{I}_s$ for any $x \in X$.

Note that any element in X is conjugate to an element in X^+ by an element in W_0. Using 2.2(h) we see that (b) is equivalent to

(c) $C\theta_x C \in \mathbf{I}_s$ for any $x \in X^+$.

This implies that

(d) If $f_{W_0}(q) \neq 0$, then $|Y_{s,q,c_0}| = 1$.

7.8. Theorem. (i). *The set Y_{s,q,c_0} is empty if and only if $\mathbf{g}_{s,q}$ is not equal to $\mathcal{N}_{s,q}$ (i.e., $\mathbf{g}_{s,q}$ contains semisimple elements).*

(ii). *If $\mathbf{g}_{s,q}$ is not equal to $\mathcal{N}_{s,q}$, then for any simple constituent L of $M_{s,0,q,1}$ we can find a nonzero nilpotent element $N \in \mathcal{N}_{s,q}$ and $\rho \in A(s,N)^\vee$ such that L is a quotient module of $M_{s,N,q,\rho}$. In particular, 5.6(d) is not true when $f_{W_0}(q) = 0$.*

Proof. (i). Suppose that $\mathbf{g}_{s,q}$ is not equal to $\mathcal{N}_{s,q}$.

According to 7.7(a-c) it is sufficient to prove that $C\theta_x C \in \mathbf{I}_s$ for any $x \in X^+$. By 7.5(a) this is equivalent to prove that

$$(7.8.1) \qquad f_{W_x}(q) \sum_{w \in W_0} x(w^{-1}(s)) \prod_{\alpha \in R^+} \frac{1 - q\alpha}{1 - \alpha}(w^{-1}(s)) = 0.$$

Note that

$$\sum_{w \in W_0} w(x \prod_{\alpha \in R^+} \frac{1 - q\alpha}{1 - \alpha}) \in \mathbb{C}[X]^{W_0}$$

is a holomorphic function on T. It is easy to check that if $\mathbf{g}_{s,q}$ is not equal to $\mathcal{N}_{s,q}$, then for each $w \in W_0$ we have

$$\prod_{\alpha \in R^+} (1 - q\alpha(w(s))) = 0.$$

When G is of classical type, we can find a sequence $t_1, t_2, ..., t_k, ...$ in T_{reg} such that $\mathbf{g}_{s,q} \neq \mathcal{N}_{s,q}$ and $\lim_{k \to \infty} t_k = s$, thus for any $w \in W_0$ we have

$$\lim_{k \to \infty} \prod_{\alpha \in R^+} \frac{1 - q\alpha(w(t_k))}{1 - \alpha(w(t_k))} = 0.$$

This implies that (7.8.1), in particular, $C\theta_x C \in \mathbf{I}_s$. Similarly using results in 6.14 we see that $C\theta_x C \in \mathbf{I}_s$ when G is of exceptional type. One direction is proved.

Now assume that $\mathbf{g}_{s,q} = \mathcal{N}_{s,q}$. Choose $(t,r) \in T \times \mathbb{C}^*$ be such that $(s,q) \sim (t,r)$ and $f_{W_0}(r) \neq 0$ (see 6.8). By (f) and 6.3 we see that $|Y_{s,q,c_0}| = 1$.

(ii). By (i) we know that $c_L \neq c_0$. Note that the nilpotent G-orbit corresponds to c_0 is $\{0\}$. Using 5.9(c-e) and 2.6(e) we get (ii).

The theorem is proved.

7.9. There are several interesting special cases. We always have $f_{W_0}(-1) = 0$.

A. Assume that $s \in T$, $q = -1$, then the following conditions are equivalent.

(a) $|Y_{s,q,c_0}| = 1$.

(b) The standard module $M_{s,q}$ is simple.

(c) $\mathbf{g}_{s,q} = \mathcal{N}_{s,q}$.

(d) $\mathbf{g}_{s,q} = \mathcal{N}_{s,q} = \{0\}$.

(e)   There is no $\alpha \in R$ such that $\alpha(s) = -1$.

(f) $tr(s, V(\delta)) \neq 0$.

By the theorem 7.8 we see that (a) and (c) are equivalent. Obviously (d) and (e) are equivalent. Since the character of $V(\delta)$ is $\delta^{-1} \prod_{\alpha \in R^+}(1 + \alpha)$ (see [Ko]), (e) and (f) are equivalent. If there is some α such that $\alpha(s) = -1$, then $\mathbf{g}_\alpha + \mathbf{g}_{-\alpha} \subseteq \mathbf{g}_{s,q}$, but the space $\mathbf{g}_\alpha + \mathbf{g}_{-\alpha}$ contains semisimple elements, thus (c) implies (e). We also have (e) implies (d) and (d) implies (c). By 5.11 we know that (d) implies (b). Note that $\dim \mathbf{H}_{s,q} = |W_0|^2$ and $\dim M_{s,q} = |W_0|$, again using 5.11 we see that (b) implies (d). We have proved these conditions are equivalent.

According to [X3] we know that (f) is equivalent to the following

(g)   In $\mathbf{H}_{-1}$ we have $C_{w_0} C_\delta = C_{w_0 \delta}$.

B. Let $q \in \mathbb{C}^*$. We assume that $f_{W_0}(q) = 0$ but for any fundamental weight $x \in X^+$ we have $f_{W_x}(q) \neq 0$. It is proved in [X3] there is a unique (up to conjugacy) semisimple element $s \in G$ such that $Y_{s,q,co} = \emptyset$. Let $w \in W_0$ be such that $f_w(q) = 0$, then by 6.9(b) and 7.8, for any element $t \in \dot{w}T$ we have $Y_{t,q,co} = \emptyset$. This implies that the elements in $\dot{w}T$ are conjugate. When q is an $(e_n + 1)$-th primitive root of 1, then $f_w(q) = 0$ if and only if w is a Coxeter element. The assertion that the elements in $\dot{w}T$ are conjugate if w is a Coxeter element was proved in [St1]. When G is of classical type, $f_w(q) = 0$ but $f_{W_x}(q) \neq 0$ (for any fundamental weight $x \in X^+$) imply that q is an (e_n+1)-th primitive root of 1. Thus $f_w(q) = 0$ if and only if w is a Coxeter element. When G is of exceptional type, $f_w(q) = 0$ but $f_{W_x}(q) \neq 0$ (for any fundamental weight $x \in X^+$) doesnot imply that q is an $(e_n + 1)$-th primitive root of 1. Thus $f_w(q) = 0$ is possible for a non-Coxeter-element w. For exceptional types we list the conjugacy classes of these elements w such that $f_w(q) = 0$ but $f_{W_x}(q) \neq 0$ for any fundamental weight $x \in X^+$. The associated types to the conjugacy classes of the elements are the same as in [C1].

Type E_8. E_8, $E_8(a_1)$, $E_8(a_2)$, $E_8(a_5)$.

Type E_7. E_7, $E_7(a_1)$.

Type E_6. E_6, $E_6(a_1)$.

Type F_4. F_4, B_4.

Type G_2. G_2, A_2.

C. Let $s \in T$ be such that $\alpha(s) = q$ for any simple root in $\alpha \in R$. We have $Y_{s,q,co} = \emptyset$ whenever $f_{W_0}(q) = 0$. This also can be proved by using 2.7.

D. If $q = 1$, the simple module in $Y_{s,q,co}$ has a simple realization which we explain now. We may assume that $s \in T$. Let L_s be the vector space over $\mathbb{C}$ with the basis $\{w(s) \mid w \in W_0\}$. There is a unique $\mathbf{H}_1 = \mathbb{C}[W]$ module structure $\cdot$ on L_s such that $u \cdot w(s) = uw(s)$ for elements u in W_0 and $\theta_x \cdot w(s) = x(w(s))$ for elements x in X. The $\mathbf{H}_1$-module L_s is obviously simple. Moreover $L_s \in Y_{s,q,co}$ since $C_{w_0} L_s \neq 0$. Note that $\dim L_s = |W_0|$ if and only if $w(s) \neq s$ whenever w is not equal to the neutral element e in W, i.e. s is a regular semisimple element.

84

8. Principal Series Representations and Induced Modules

Let (W, S) be the extended affine Weyl group associated to a simple complex algebraic group G and let $\mathbf{H}_q$ ($q \in \mathbb{C}^*$) be the Hecke algebras of (W, S) defined in 5.6. Matsumoto introduced certain principal series representations in [M, 4.1.5, p.87] and Lusztig introduced another model for principal series representations [L2, 8.11, p.146]. Matsumoto's model is elementary but the intertwining operators have poles and hence are not everywhere defined [M, 4.3.2]. Lusztig's model is less elementary but the intertwining operators are everywhere defined, actually they are isomorphisms, see [L2, Prop. 2.8].

In this chapter we define a model for principal series representations for each (generalized) two-sided cell of W. Matsumoto's model then is the model corresponding to the highest two-sided cell Ω of W (with respect to the partial order $\underset{LR}{\leq}$ on $\mathrm{Cell}(W)$) and Lusztig's model is the model corresponding to the lowest two-sided cell c_0 of W. All the models are induced modules in some sense.

8.1. To be convenient we recall some notions and notations. We fix a maximal torus T of G and let $X = \mathrm{Hom}(T, \mathbb{C}^*)$ be the character group of T. The Weyl group W_0 of G acts on X and the semi-direct product $W_0 \ltimes X$ is just the extended affine Weyl group W. Moreover W is isomorphic to $\Omega \ltimes W_a$ for some finite abelian group Ω and some affine Weyl group W_a.

Recall that in 5.6 we have defined the Hecke algebra $\mathbf{H}_q$ (over $\mathbb{C}$) of W for each $q \in \mathbb{C}^*$. We keep the notaitons $T_w, C_w, \theta_x, U_x, \dots$. Denote Θ_q the subalgebra of $\mathbf{H}_q$ generated by $\theta_x, x \in X$. For each element s in the maximal torus T of G we denote $\mathbb{C}_s$ the one-dimensional Θ_q-module on which θ_x ($x \in X$) acts by scalar $x(s)$.

Let c be a two-sided cell of W and let u be an element in c. The elements $C_w, w \underset{LR}{\leq} u$ span a two-sided ideal $\mathbf{H}_q^{\leq c}$ of $\mathbf{H}_q$ and the elements $C_w, w \underset{LR}{\leq} u, w \notin c$ span a two-sided ideal $\mathbf{H}_q^{<c}$ of $\mathbf{H}_q$. Since $\mathbf{H}_q$ is a free right Θ_q-module of rank $|W_0|$ and Θ_q is a noetherian ring, both $\mathbf{H}_q^{\leq c}$ and $\mathbf{H}_q^{<c}$ are finitely generated right Θ_q-modules, where the actions of Θ_q are defined through multiplication in $\mathbf{H}_q$.

Let $s \in T$ and let c be a two-sided cell of W. Set

$$M_{s,c} := \mathbf{H}_q^{\leq c} \otimes_{\Theta_q} \mathbb{C}_s.$$

Naturally $M_{s,c}$ is an $\mathbf{H}_q$-module:

$$h(h' \otimes a) = hh' \otimes a.$$

When c is the highest two-sided cell Ω of W, the module $M_{s,c}$ is just the principal series representation constructed in [M, 4.1.5]. When c is the lowest two-sided cell c_0 of W, the module $M_{s,c}$ is just the module $\mathcal{M}_s$ constructed in [L2, 8.11]. Let c'

be a two-sided cell of W such that $c' \underset{LR}{\leq} c$, the embedding $\mathbf{H}_q^{\leq c'} \hookrightarrow \mathbf{H}_q^{\leq c}$ gives rise to a homomorphism of $\mathbf{H}_q$-module

$$A_{c',c} : M_{s,c'} \to M_{s,c}.$$

When $c' = c_0$ and $c = \Omega$, the homomorphism $A_{c',c}$ is just that defined in [L2, 8.11]. We conjecture that $\dim M_{s,c} = |W_0|$ for any $s \in T$ and any two-sided cell c of W.

8.2. Lemma. *Let $s \in T$. Then M_{s,c_0} is isomorphic to $M_{w(s),c_0}$ for any w in W_0.*

Proof. Let $\mathbf{I}$ be the left ideal of $\mathbf{H}_q$ generated by $C = \sum_{w \in W_0} T_w$ and let $\mathbf{I}_s$ be the left ideal of $\mathbf{H}_q$ generated by $(U_x - tr(s, V(x)))C$, $x \in X^+$. The embedding $\mathbf{I} \hookrightarrow \mathbf{H}_q^{\leq c_0}$ gives rise to an $\mathbf{H}_q$-homomorphism $\varphi_s : \mathbf{I} \to M_{s,c_0}$, $h \to h \otimes 1$. Obviously φ_s is surjective and sends $\mathbf{I}_s$ to zero. Hence φ_s induces a surjective homomorphism of $\mathbf{H}_q$-module $\psi_s : \mathbf{I}/\mathbf{I}_s \to M_{s,c_0}$. Both $\mathbf{I}/\mathbf{I}_s$ and M_{s,c_0} have the same dimension $|W_0|$, so ψ_s is an isomorphism of $\mathbf{H}_q$-module. Obviously $\mathbf{I}/\mathbf{I}_s = \mathbf{I}/\mathbf{I}_{w(s)}$ for any w in W_0, the lemma follows.

Remark: Assume that G is simply connected. By the above proof and Lemma 7.6, M_{s,c_0} is isomorphic to the standard module $M_{s,0,q,1}$.

8.3. Let $s \in T$ and c a two-sided cell of W. Assume that r is a simple reflection in W_0 and α is the corresponding simple root. We may define the intertwining operators between $M_{s,c}$ and $M_{r(s),c}$ as follows (compare with [M, 4.3.2] and [R1, section 2]).

When $\alpha(s) = 1$, then $r(s) = s$, we need to do nothing. Now assume that $\alpha(s) \neq 1$. Set

$$\xi_{r,s} = T_r + \frac{q-1}{\alpha(s)-1}, \qquad \xi'_{r,s} = T_r - \frac{\alpha(s)(q-1)}{\alpha(s)-1}.$$

Define

$$\tilde{\xi}_{r,s} : \mathbf{H}_q^{\leq c} \to \mathbf{H}_q^{\leq c}, \quad h \to h\xi_{r,s},$$

$$\tilde{\xi}'_{r,s} : \mathbf{H}_q^{\leq c} \to \mathbf{H}_q^{\leq c}, \quad h \to h\xi'_{r,s},$$

which induce two homomorphisms of $\mathbf{H}_q$-module:

$$\tilde{\xi}_{r,s} : \mathbf{H}_q^{\leq c} \to M_{r(s),c} = \mathbf{H}_q^{\leq c} \otimes_{\Theta_q} \mathbb{C}_{r(s)}, \quad h \to h\xi_{r,s} \otimes 1,$$

$$\tilde{\xi}'_{r,s} : \mathbf{H}_q^{\leq c} \to M_{s,c} = \mathbf{H}_q^{\leq c} \otimes_{\Theta_q} \mathbb{C}_s, \quad h \to h\xi'_{r,s} \otimes 1.$$

Let $x \in X$ be such that $\langle x, \alpha^\vee \rangle = n \geq 0$, then

$$\tilde{\xi}_{r,s}(h\theta_x) = h\theta_x(T_r + \frac{q-1}{\alpha(s)-1}) \otimes 1$$

$$= h(T_r\theta_{r(x)} + (q-1)\theta_x \frac{\theta_\alpha^{-n}-1}{\theta_\alpha^{-1}-1} + \frac{q-1}{\alpha(s)-1}\theta_x) \otimes 1$$

$$= h(T_r x(s) + \frac{q-1}{\alpha(s)-1}x(s)) \otimes 1$$

$$= \tilde{\xi}_{r,s}(x(s)h).$$

86

So $\tilde{\xi}_{r,s}$ induces an $\mathbf{H}_q$-homomorphism

$$\eta_{r,s} : M_{s,c} = \mathbf{H}_q^{\le c} \otimes_{\Theta_q} \mathbb{C}_s \to M_{r(s),c} = \mathbf{H}_q^{\le c} \otimes_{\Theta_q} \mathbb{C}_{r(s)}, \quad h \otimes 1 \to h\xi_{r,s} \otimes 1.$$

Similarly $\tilde{\xi}'_{r,s}$ induces an $\mathbf{H}_q$-homomorphism

$$\eta'_{r,s} : M_{r(s),c} = \mathbf{H}_q^{\le c} \otimes_{\Theta_q} \mathbb{C}_{r(s)} \to M_{s,c} = \mathbf{H}_q^{\le c} \otimes_{\Theta_q} \mathbb{C}_s, \quad h \otimes 1 \to h\xi'_{r,s} \otimes 1.$$

When $\alpha(s) \ne q^{\pm 1}$, neither $\dfrac{q-1}{\alpha(s)-1}$ nor $-\dfrac{\alpha(s)(q-1)}{\alpha(s)-1}$ is equal to 1 or $-q$. So both $\xi_{r,s}$ and $\xi'_{r,s}$ are invertible in $\mathbf{H}_q$. Actually, if $a \ne 1, -q$, then

$$(T_r + a)\frac{T_r + 1 - q - a}{(1-a)(a+q)} = 1.$$

Therefore both $\eta_{r,s}$ and $\eta'_{r,s}$ are isomorphisms.

When $\alpha(s) = q$, then $\xi_{r,s} = T_r + 1$ and $\xi'_{r,s} = T_r - q$. So $\eta_{r,s}\eta'_{r,s} = \eta'_{r,s}\eta_{r,s} = 0$. If $q \ne -1$, then

$$\mathbf{H}_q^{\le c} = \mathbf{H}_q^{\le c}\xi_{r,s} \oplus \mathbf{H}_q^{\le c}\xi'_{r,s}.$$

Let A_s and $A_{r(s)}$ be the images in $M_{s,c}$ and in $M_{r(s),c}$ of $\mathbf{H}_q^{\le c}\xi_{r,s}$ respectively and let B_s and $B_{r(s)}$ be the images in $M_{s,c}$ and in $M_{r(s),c}$ of $\mathbf{H}_q^{\le c}\xi'_{r,s}$ respectively. Then $M_{s,c} = A_s + B_s$ and $M_{r(s),c} = A_{r(s)} + B_{r(s)}$. Obviously we have $\eta_{r,s}(B_s) = 0$, $\eta_{r,s}(A_s) = A_{r(s)}$; and $\eta'_{r,s}(A_{r(s)}) = 0$, $\eta'_{r,s}(B_{r(s)}) = B_s$. If $q = -1$, then $\xi_{r,s} = \xi'_{r,s} = T_r + 1$. Let A_s and $A_{r(s)}$ be the images in $M_{s,c}$ and $M_{r(s),c}$ of $\mathbf{H}_q^{\le c}\xi_{r,s}$ respectively, then $\eta_{r,s}(A_s) = 0$, $\eta_{r,s}(M_{s,c}) = A_{r(s)}$; and $\eta'_{r,s}(A_{r(s)}) = 0$, $\eta'_{r,s}(M_{r(s),c}) = A_s$. Similarly we may discuss the case $\alpha(s) = q^{-1}$.

Assume that $c = \Omega$. Then $\mathbf{H}_q^{\le c}$ is a free right Θ_q-module. Let $\mathbf{H}'_q$ be the subalgebra of $\mathbf{H}_q$ generated by $T_w, w \in W_0$. If $\alpha(s) = q^{\pm 1}$, then the kernel of $\eta_{r,s}$ is $\mathbf{H}'_q\xi'_{r,s} \otimes 1$ and the kernel of $\eta'_{r,s}$ is $\mathbf{H}'_q\xi_{r,s} \otimes 1$. Both kernels are of dimension $|W_0|/2$ (cf. [R1, section 2]), in particular, $M_{s,c}$ and $M_{r(s),c}$ have the same composition factors. Therefore when $c = \Omega$, $M_{s,c}$ and $M_{w(s),c}$ have the same composition factors for any w in W_0 [R1, Prop. 2.3].

8.4. Conjecture. *Let c, c' be two-sided cells of W and $s \in T$.*

(i). *For any element w in W_0, $M_{s,c}$ and $M_{w(s),c}$ have the same composition factors.*

(ii). *$M_{s,c}$ and $M_{s,c'}$ have the same composition factors.*

8.5. More generally, let V be an $(\mathbf{H}_q, \Theta_q)$-module (i.e. V is a left $\mathbf{H}_q$-module and a right Θ_q-module and the actions commute: $h(v\theta_x) = h(v)\theta$ for all h in $\mathbf{H}_q$, v in V and θ in Θ_q). For any s in T, $V_s := V \otimes_{\Theta_q} \mathbb{C}_s$ is naturally an $\mathbf{H}_q$-module: $h(v \otimes a) = hv \otimes a$. An interesting case is the following.

Let c be a two-sided cell of W and let $s \in T$. Obviously $E_c := \mathbf{H}_q^{\le c}/\mathbf{H}_q^{< c}$ is an $(\mathbf{H}_q, \Theta_q)$-module. Therefore

$$E_{s,c} := E_c \otimes_{\Theta_q} \mathbb{C}_s$$

87

is an $\mathbf{H}_q$-module of finite dimension. Note that E_{s,c_0} is just M_{s,c_0}.

Let $\mathbf{g}$ be the Lie algebra of G and let $\mathcal{N}_c$ be the nilpotent class of $\mathbf{g}$ corresponding to the two-sided cell c.

8.6. Conjecture. *Suppose that G is simply connected.*

(a). *Assume that $E_{s,c}$ is not zero. Then*

    (i). *There exists an element N in $\mathcal{N}_c$ such that $s.N = qN$.*

    (ii). *Let N be as in (i), then as $\mathbf{H}_q$-modules, $E_{s,c}$ is isomorphic to $\mathbf{K}(\mathcal{B}_N^s) = H_*(\mathcal{B}_N^s)$ (see 5.6).*

(b). *Assume that $E_{s,c}$ is zero, then it is impossible to find an element N in $\mathcal{N}_c$ such that $s.N = qN$.*

8.7. We may further construct an $\mathbf{H}_q$-module $E_{s,c,p}$ for any s in T, two-sided cell c of W and p in $\mathbb{C}^*$. Recall that we have a right $\mathbf{H}_p$-module structure on $\mathbf{H}_q^{\leq c}$ and on $\mathbf{H}_q^{<c}$ defined as follows (cf. [L11, 9.1, p.274]. We define $C_w T_\omega = C_{w\omega}$ for ω in Ω and

$$C_w T_r = \begin{cases} pC_w, & \text{if } wr \leq w, \\ p^{\frac{1}{2}}(C_{wr} + \displaystyle\sum_{\substack{y \prec w \\ yr \leq y}} \mu(y,w)C_y) - C_w, & \text{if } wr \geq w \end{cases}$$

for r in S. The left $\mathbf{H}_q$-module structure on $\mathbf{H}_q^{\leq c}$ or $\mathbf{H}_q^{<c}$ usually doesnot commute with the right $\mathbf{H}_p$-module structure on $\mathbf{H}_q^{\leq c}$ or $\mathbf{H}_q^{<c}$. However, the left $\mathbf{H}_q$-module structure on $E_c = \mathbf{H}_q^{\leq c}/\mathbf{H}_q^{<c}$ commutes with the right $\mathbf{H}_p$-module structure on E_c [L11, Theorem 9.2].

Let Θ_p be the subalgebra of $\mathbf{H}_p$ generated by $\theta_x, x \in X$ and for s in T we denote also $\mathbb{C}_s$ the corresponding one-dimensional Θ_p-module. Thus we have a natural $\mathbf{H}_q$-module structure on

$$E_{s,c,p} := E_c \otimes_{\Theta_p} \mathbb{C}_s, \qquad h(t \otimes a) = ht \otimes a.$$

Noting that $\mathbf{H}_q$ is a free right Θ_p-module of rank $|W_0|$ and Θ_p is a noetherian ring, we see that $\mathbf{H}_q^{\leq c}$ is a finitely generated right Θ_p-module. Therefore $E_{s,c,p}$ is an $\mathbf{H}_q$-module of finite dimension. Let L be an irreducible $\mathbf{H}_p$-module, then $L_{s,c} := E_c \otimes_{\mathbf{H}_p} L$ is an $\mathbf{H}_q$-module of finite dimension. Maybe the $\mathbf{H}_q$-modules $E_{s,c,p}$ and $L_{s,c}$ are interesting in understanding the relations between representations of $\mathbf{H}_p$ and representations of $\mathbf{H}_q$.

Taking p to be 1, then $\mathbf{H}_p$ is the group algebra $\mathbb{C}[W]$ of W. Thus for each w in W we get an $\mathbf{H}_q$-homomorphism

$$B_w : E_c \to E_c, \quad t \to tw.$$

For w, u in W we have $B_w B_u = B_{uw}$. When c is the lowest two-sided cell c_0, E_c is equal to $\mathbf{H}_q^{\leq c_0}$. In this case are there any relations between the isomorphism B_w and the isomorphism θ_w define in [L2, Prop. 2.8]?

8.8. Let S_1 be a subset of S_0 and let $\mathbf{H}_q^{(1)}$ be the subalgebra of $\mathbf{H}_q$ generated by T_r, θ_x, $r \in S_1, x \in X$. Obviously $\mathbf{H}_q$ is a free right $\mathbf{H}_q^{(1)}$-module of rank $|W_0|/|W_1|$, where W_1 is the subgroup of W_0 generated by S_1 and the action of $\mathbf{H}_q^{(1)}$ is defined through the multiplication in $\mathbf{H}_q$. Let M_1 be an $\mathbf{H}_q^{(1)}$-module and define

$$\text{ind} M_1 := \mathbf{H}_q \otimes_{\mathbf{H}_q^{(1)}} M_1.$$

When M_1 is of finite dimension, so is $\text{ind} M_1$. The induced modules $\text{ind} M_1$ play an interesting role in the work [KL4]. We actually can define the following induced modules for each two-sided cell c of W:

$$\text{ind}^{\leq c} M_1 := \mathbf{H}_q^{\leq c} \otimes_{\mathbf{H}_q^{(1)}} M_1,$$

$$\text{ind}^c M_1 := \mathbf{H}_q^{\leq c} / \mathbf{H}_q^{<c} \otimes_{\mathbf{H}_q^{(1)}} M_1.$$

It seems not easy to determine whether these $\mathbf{H}_q$-modules $\text{ind}^{\leq c} M_1$ and $\text{ind}^c M_1$ are interesting.

8.9. The previous discuss may carry to affine Hecke algebras with two parameters. In the rest part of the chapter G will be one of the types B_n, C_n, F_4, G_2. We keep the notations in 2.8-2.9. Assume that we have a total order on the abelian group $\{\mathbf{u}^i \mathbf{v}^j \mid i, j \in \mathbb{Z}\}$ which is compatible with the multiplication. Assume further that $\mathbf{u} > 1$, $\mathbf{v} > 1$, then we can define generalized cells of W (with respect to the given order and $\varphi : S \to \{\mathbf{u}, \mathbf{v}\}$). Recall that for each $w \in W$ we have an associated element $C_w \in \check{H}$ (see 1.14 (a)). The image in $\tilde{H}_{a,b}$ $(a, b \in \mathbb{C}^*)$ of C_w will denote by the same notation. The subalgebra of $\tilde{H}_{a,b}$ generated by $\theta_x, x \in X$ will denote by $\Theta_{a,b}$. For each element s in the maximal torus T of G we again denote $\mathbb{C}_s$ the one-dimensional $\Theta_{a,b}$-module on which θ_x $(x \in X)$ acts by scalar $x(s)$.

Let c be a generalized two-sided cell of W and let u be an element in c. The elements $C_w, w \underset{LR,\varphi}{\leq} u$ span a two-sided ideal $\tilde{H}_{a,b}^{\leq c}$ and the elements $C_w, w \underset{LR,\varphi}{\leq} u, w \notin c$ span a two-sided ideal $\tilde{H}_{a,b}^{<c}$. Since $\tilde{H}_{a,b}$ is a free right $\Theta_{a,b}$-module of rank $|W_0|$ and $\Theta_{a,b}$ is a noetherian ring, both $\tilde{H}_{a,b}^{\leq c}$ and $\tilde{H}_{a,b}^{<c}$ are finitely generated right $\Theta_{a,b}$-modules, where the actions of $\Theta_{a,b}$ are defined through multiplication in $\tilde{H}_{a,b}$.

Let $s \in T$ and let c be a generalized two-sided cell of W. Set

$$M_{s,c} := \tilde{H}_{a,b}^{\leq c} \otimes_{\Theta_{a,b}} \mathbb{C}_s.$$

Naturally $M_{s,c}$ is an $\tilde{H}_{a,b}$-module:

$$h(h' \otimes a) = hh' \otimes a.$$

When c is the highest generalized two-sided cell Ω of W, the module $M_{s,c}$ is just the principal series representation constructed in [M, 4.1.5]. When c is the lowest generalized two-sided cell c_0 of W (cf. 3.22), the module $M_{s,c}$ is similar to the module

$\mathcal{M}_s$ constructed in [L2, 8.11]. Let c' be a two-sided cell of W such that $c' \underset{LR,\varphi}{\leq} c$, the embedding $\tilde{\mathbf{H}}_{a,b}^{\leq c'} \hookrightarrow \tilde{\mathbf{H}}_{a,b}^{\leq c}$ gives rise to a homomorphism of $\tilde{\mathbf{H}}_{a,b}$-module

$$A_{c',c} : M_{s,c'} \to M_{s,c}.$$

When $c' = c_0$ and $c = \Omega$, the homomorphism $A_{c',c}$ is similar to that defined in [L2, 8.11]. We conjecture that $\dim M_{s,c} = |W_0|$ for any generalized two-sided cell c of W (with respect to the given order and φ).

Let $\mathbf{I}$ be the left ideal of $\tilde{\mathbf{H}}_{a,b}$ generated by $C = \sum_{w \in W_0} T_w$ and let $\mathbf{I}_s$ be the left ideal of $\tilde{\mathbf{H}}_{a,b}$ generated by $(U_x - tr(s,V(x))C$, $x \in X^+$. The embedding $\mathbf{I} \hookrightarrow \tilde{\mathbf{H}}_{a,b}^{\leq c_0}$ gives rise to an $\tilde{\mathbf{H}}_{a,b}$-homomorphism $\varphi_s : \mathbf{I} \to M_{s,c_0}$. Obviously φ_s is surjective and sends $\mathbf{I}_s$ to zero. Hence φ_s induces a surjective homomorphism of $\tilde{\mathbf{H}}_{a,b}$-module $\psi_s : \mathbf{I}/\mathbf{I}_s \to M_{s,c_0}$. Since both $\mathbf{I}/\mathbf{I}_s$ and M_{s,c_0} have dimension $|W_0|$, ψ_s is an isomorphism of $\tilde{\mathbf{H}}_{a,b}$-module. Since $\mathbf{I}_s$ is equal to $\mathbf{I}_{w(s)}$ for any w in W_0, we get

(a). As $\tilde{\mathbf{H}}_{a,b}$-modules, M_{s,c_0} is isomorphic to $M_{w(s),c_0}$ for any w in W_0.

8.10. To be convenient we set $q_r = a$ if $r \in S'$ and $q_r = b$ if $r \in S''$. Let $s \in T$ and c a generalized two-sided cell of W. Assume that r is a simple reflection in W_0 and α is the corresponding simple root. We may define the intertwining operators between $M_{s,c}$ and $M_{r(s),c}$ as follows (compare with [M, 4.3.2] and [R1, section 2]).

When $\alpha(s) = 1$, then $r(s) = s$, we need to do nothing. Now assume that $\alpha(s) \neq 1$. Set

$$\xi_{r,s} = T_r + \frac{q_r - 1}{\alpha(s) - 1}, \qquad \xi'_{r,s} = T_r - \frac{\alpha(s)(q_r - 1)}{\alpha(s) - 1}.$$

Define

$$\tilde{\xi}_{r,s} : \tilde{\mathbf{H}}_{a,b}^{\leq c} \to \tilde{\mathbf{H}}_{a,b}^{\leq c}, \quad h \to h\xi_{r,s},$$

$$\tilde{\xi}'_{r,s} : \tilde{\mathbf{H}}_{a,b}^{\leq c} \to \tilde{\mathbf{H}}_{a,b}^{\leq c}, \quad h \to h\xi'_{r,s},$$

which induce two homomorphisms of $\tilde{\mathbf{H}}_{a,b}$-module:

$$\tilde{\xi}_{r,s} : \tilde{\mathbf{H}}_{a,b}^{\leq c} \to M_{r(s),c} = \tilde{\mathbf{H}}_{a,b}^{\leq c} \otimes_{\Theta_{a,b}} \mathbb{C}_{r(s)}, \quad h \to h\xi_{r,s} \otimes 1,$$

$$\tilde{\xi}'_{r,s} : \tilde{\mathbf{H}}_{a,b}^{\leq c} \to M_{s,c} = \tilde{\mathbf{H}}_{a,b}^{\leq c} \otimes_{\Theta_{a,b}} \mathbb{C}_s, \quad h \to h\xi'_{r,s} \otimes 1.$$

Let $x \in X$ be such that $\langle x, \alpha^\vee \rangle = n \geq 0$, then

$$\tilde{\xi}_{r,s}(h\theta_x) = h\theta_x(T_r + \frac{q_r - 1}{\alpha(s) - 1}) \otimes 1$$

$$= h(T_r\theta_{r(x)} + (q_r - 1)\theta_x \frac{\theta_\alpha^{-n} - 1}{\theta_\alpha^{-1} - 1} + \frac{q_r - 1}{\alpha(s) - 1}\theta_x) \otimes 1$$

$$= h(T_r x(s) + \frac{q_r - 1}{\alpha(s) - 1} x(s)) \otimes 1$$

$$= \tilde{\xi}_{r,s}(x(s)h).$$

90

So $\tilde{\xi}_{r,s}$ induces an $\tilde{\mathbf{H}}_{a,b}$-homomorphism

$$\eta_{r,s} : M_{s,c} = \tilde{\mathbf{H}}_{a,b}^{\leq c} \otimes_{\Theta_{a,b}} \mathbb{C}_s \to M_{r(s),c} = \tilde{\mathbf{H}}_{a,b}^{\leq c} \otimes_{\Theta_{a,b}} \mathbb{C}_{r(s)}, \quad h \otimes 1 \to h\xi_{r,s} \otimes 1.$$

Similarly $\tilde{\xi}'_{r,s}$ induces an $\tilde{\mathbf{H}}_{a,b}$-homomorphism

$$\eta'_{r,s} : M_{r(s),c} = \tilde{\mathbf{H}}_{a,b}^{\leq c} \otimes_{\Theta_{a,b}} \mathbb{C}_{r(s)} \to M_{s,c} = \tilde{\mathbf{H}}_{a,b}^{\leq c} \otimes_{\Theta_{a,b}} \mathbb{C}_s, \quad h \otimes 1 \to h\xi'_{r,s} \otimes 1.$$

When $\alpha(s) \neq q_r^{\pm 1}$, neither $\dfrac{q_r - 1}{\alpha(s) - 1}$ nor $-\dfrac{\alpha(s)(q_r - 1)}{\alpha(s) - 1}$ is equal to 1 or $-q_r$. So both $\xi_{r,s}$ and $\xi'_{r,s}$ are invertible in $\tilde{\mathbf{H}}_{a,b}$. Actually, if $a \neq 1, -q_r$, then

$$(T_r + a)\frac{T_r + 1 - q_r - a}{(1 - a)(a + q_r)} = 1.$$

Therefore both $\eta_{r,s}$ and $\eta'_{r,s}$ are isomorphisms.

When $\alpha(s) = q_r$, then $\xi_{r,s} = T_r + 1$ and $\xi'_{r,s} = T_r - q_r$. So $\eta_{r,s}\eta'_{r,s} = \eta'_{r,s}\eta_{r,s} = 0$. If $q_r \neq -1$, then

$$\tilde{\mathbf{H}}_{a,b}^{\leq c} = \tilde{\mathbf{H}}_{a,b}^{\leq c}\xi_{r,s} \oplus \tilde{\mathbf{H}}_{a,b}^{\leq c}\xi'_{r,s}.$$

Let A_s and $A_{r(s)}$ be the images in $M_{s,c}$ and in $M_{r(s),c}$ of $\tilde{\mathbf{H}}_{a,b}^{\leq c}\xi_{r,s}$ respectively and let B_s and $B_{r(s)}$ be the images in $M_{s,c}$ and in $M_{r(s),c}$ of $\tilde{\mathbf{H}}_{a,b}^{\leq c}\xi'_{r,s}$ respectively. Then $M_{s,c} = A_s + B_s$ and $M_{r(s),c} = A_{r(s)} + B_{r(s)}$. Obviously we have $\eta_{r,s}(B_s) = 0$, $\eta_{r,s}(A_s) = A_{r(s)}$; and $\eta'_{r,s}(A_{r(s)}) = 0$, $\eta'_{r,s}(B_{r(s)}) = B_s$. If $q_r = -1$, then $\xi_{r,s} = \xi'_{r,s} = T_r + 1$. Let A_s and $A_{r(s)}$ be the images in $M_{s,c}$ and $M_{r(s),c}$ of $\tilde{\mathbf{H}}_{a,b}^{\leq c}\xi_{r,s}$ respectively, then $\eta_{r,s}(A_s) = 0$, $\eta_{r,s}(M_{s,c}) = A_{r(s)}$; and $\eta'_{r,s}(A_{r(s)}) = 0$, $\eta'_{r,s}(M_{s,c}) = A_s$. Similarly we may discuss the case $\alpha(s) = q_r^{-1}$.

Assume that $c = \Omega$. Then $\tilde{\mathbf{H}}_{a,b}^{\leq c}$ is a free right $\Theta_{a,b}$-module. Let $\tilde{\mathbf{H}}'_{a,b}$ be the subalgebra of $\tilde{\mathbf{H}}_{a,b}$ generated by $T_w, w \in W_0$. If $\alpha(s) = q_r^{\pm 1}$, then the kernel of $\eta_{r,s}$ is $\tilde{\mathbf{H}}'_{a,b}\xi'_{r,s} \otimes 1$ and the kernel of $\eta'_{r,s}$ is $\tilde{\mathbf{H}}'_{a,b}\xi_{r,s} \otimes 1$. Both kernels are of dimension $|W_0|/2$ (cf. [R1, section 2]), in particular, $M_{s,c}$ and $M_{r(s),c}$ have the. same composition factors. Therefore when $c = \Omega$, $M_{s,c}$ and $M_{w(s),c}$ have the same composition factors for any w in W_0 (compare with [R1, Prop. 2.3]).

8.11 Conjecture. *Let c, c' be generalized two-sided cells of W and $s \in T$.*

(i). *For each element w in W_0, the $\tilde{\mathbf{H}}_{a,b}$-modules $M_{s,c}$ and $M_{w(s),c}$ have the same composition factors.*

(ii). *The $\tilde{\mathbf{H}}_{a,b}$-modules $M_{s,c}$ and $M_{s,c'}$ have the same composition factors.*

8.12. More generally, let V be an $(\tilde{\mathbf{H}}_{a,b}, \Theta_{a,b})$-module (i.e. V is a left $\tilde{\mathbf{H}}_{a,b}$-module and a right $\Theta_{a,b}$-module and the actions commute: $h(v\theta_x) = h(v)\theta$ for all h in $\tilde{\mathbf{H}}_{a,b}$, v in V and θ in $\Theta_{a,b}$). For any s in T, $V_s := V \otimes_{\Theta_{a,b}} \mathbb{C}_s$ is naturally an $\tilde{\mathbf{H}}_{a,b}$-module: $h(v \otimes a) = hv \otimes a$. An interesting case is the following.

91

Let c be a generalized two-sided cell of W and let $s \in T$. Obviously $E_c := \tilde{\mathbf{H}}_{a,b}^{\leq c}/\tilde{\mathbf{H}}_{a,b}^{<c}$ is an $(\tilde{\mathbf{H}}_{a,b}, \Theta_{a,b})$-module. Therefore

$$E_{s,c} := E_c \otimes_{\Theta_{a,b}} \mathbb{C}_s$$

is an $\tilde{\mathbf{H}}_{a,b}$-module of finite dimension. Note that E_{s,c_0} is just M_{s,c_0}.

Let S_1 be a subset of S_0 and let $\tilde{\mathbf{H}}_{a,b}^{(1)}$ be the subalgebra of $\tilde{\mathbf{H}}_{a,b}$ generated by T_r, θ_x, $r \in S_1, x \in X$. Obviously $\tilde{\mathbf{H}}_{a,b}$ is a free right $\tilde{\mathbf{H}}_{a,b}^{(1)}$-module of rank $|W_0|/|W_1|$, where W_1 is the subgroup of W_0 generated by S_1 and the action of $\tilde{\mathbf{H}}_{a,b}^{(1)}$ is defined through multiplication in $\tilde{\mathbf{H}}_{a,b}$. Let M_1 be an $\tilde{\mathbf{H}}_{a,b}^{(1)}$-module and let c be a generalized two-sided cell of W. Define

$$\text{ind}^{\leq c} M_1 := \tilde{\mathbf{H}}_{a,b}^{\leq c} \otimes_{\tilde{\mathbf{H}}_{a,b}^{(1)}} M_1,$$

$$\text{ind}^c M_1 := \tilde{\mathbf{H}}_{a,b}^{\leq c}/\tilde{\mathbf{H}}_{a,b}^{<c} \otimes_{\tilde{\mathbf{H}}_{a,b}^{(1)}} M_1.$$

When c is the highest generalized two-sided cell Ω of W, $\tilde{\mathbf{H}}_{a,b}^{\leq c} = \tilde{\mathbf{H}}_{a,b}$. When c is the lowest generalized two-sided cell c_0 of W, $\tilde{\mathbf{H}}_{a,b}^{\leq c}$ is also clear (cf. Theorem 3.22). In these cases it is possible to discuss the induced modules $\text{ind}^{\leq c} M_1$ and $\text{ind}^c M_1$. In general it is hard to discuss these induced modules since we know too little about the generalized two-sided cells c.

8.13. Let $a', b' \in \mathbb{C}^*$. We may have a right $\tilde{\mathbf{H}}_{a',b'}$-module structure on $\tilde{\mathbf{H}}_{a,b}^{\leq c}$ and on $\tilde{\mathbf{H}}_{a,b}^{<c}$ as follows (cf. [L11, 9.1, p.274]. We define $C_w T_\omega = C_{w\omega}$ for ω in Ω and

$$C_w T_r = \begin{cases} p_r C_w, & \text{if } wr \leq w, \\ p_r^{\frac{1}{2}}(C_{wr} + \sum_{\substack{y \\ yr \leq y}} m_{y,w}^r C_y) - C_w, & \text{if } wr \geq w \end{cases}$$

for r in S, where $p_r = a'$ if $r \in S'$ and $p_r = b'$ if $r \in S''$, and $m_{y,w}^r$ is the value of $M_{y^{-1},z^{-1}}^r$ (see 1.14) at (a', b'). The left $\tilde{\mathbf{H}}_{a,b}$-module structure on $\tilde{\mathbf{H}}_{a,b}^{\leq c}$ or $\tilde{\mathbf{H}}_{a,b}^{<c}$ usually doesnot commute with the right $\tilde{\mathbf{H}}_{a',b'}$-module structure on $\tilde{\mathbf{H}}_{a,b}^{\leq c}$ or $\tilde{\mathbf{H}}_{a,b}^{<c}$. The question is whether the left $\tilde{\mathbf{H}}_{a,b}$-module structure on $E_c = \tilde{\mathbf{H}}_{a,b}^{\leq c}/\tilde{\mathbf{H}}_{a,b}^{<c}$ commutes with the right $\tilde{\mathbf{H}}_{a',b'}$-module structure on E_c? (cf. [L11, Theorem 9.2]). We conjecture that on E_c the $\tilde{\mathbf{H}}_{a,b}$-action commutes with the $\tilde{\mathbf{H}}_{a',b'}$-action when c is the lowest generalized two-sided cell c_0.

9. Isogenous Affine Hecke Algebras

Let $W = \Omega \ltimes W_{\text{aff}}$ and $W' = \Omega' \ltimes W'_{\text{aff}}$ be two extended affine Weyl groups. We say that W and W' are isogenous if W_{aff} and W'_{aff} are isomorphic affine Weyl groups. The Hecke algerbas (over the same ring and with the same set of parameters) of isogenous extended affine groups are called isogenous. In this chapter we will discuss some relations among isogenous extended affine Weyl groups and isogenous affine Hecke algebras, which seem interesting. All algebraic groups will be complex algebraic groups.

9.1. Let G be a connected reductive group and T a maximal torus of G. Denote $G' = (G, G)$ the derived group of G and set $T' = T \cap G'$. Then T' is a maximal torus of G'. The following facts (a-e) are well known.

(a) The injection $T' \hookrightarrow T$ induces a surjective homomorphism

$$f : X = \text{Hom}(T, \mathbb{C}^*) \to X' = \text{Hom}(T', \mathbb{C}^*).$$

(b) The Weyl group $W_0 = N_G(T)/T$ of G is naturally isomorphic to the Weyl group $W'_0 = N_{G'}(T')/T'$ of G'.

(c) If we identify W_0 with W'_0, then the homomorphism f in (a) is W_0-equivariant.

Therefore we have

Assertion A. *The extended affine Weyl groups $W = W_0 \ltimes X$ and $W' = W'_0 \ltimes X'$ are isogenous.*

Actually, let $R \subset X$ be the root system of W_0, then $f(R)$ is the root system of W'_0. Now using (b) we see that the affine Weyl groups associated to W and to W' are isomorphic. That is, W and W' are isogenous.

Let Z_G be the center of G. Then Z_G is connected and $Z_G \cap T'$ is finite. Moreover $Z_G T' = T$. It is easy to check the following

(d) $\ker f = \{x \in X \mid w(x) = x \text{ for all } w \in W_0\}$ and $K := \ker f$ is naturally isomorphic to $\text{Hom}(Z_G/(Z_G \cap T'), \mathbb{C}^*)$.

It is known that X, X' and K are free abelian groups of finite rank. Let $y_1, ..., y_m$ be elements in X such that $f(y_1), ..., f(y_m)$ form a basis of X' and let $z_1, ..., z_k$ be a basis of K, then $y_1, ..., y_m, z_1, ..., z_k$ form a basis of X. So we have

(e) X is isomorphic to the direct product $K \times X'$.

From (c-e) we get

(f) W is isomorphic to $K \times W'$.

Denote $\mathbf{H}_q$ (resp. $\mathbf{H}'_q$) the Hecke algebra over $\mathbb{C}$ of W (resp. of W') with parameter $q \in \mathbb{C}^*$. Then we have

Assertion B. *As $\mathbb{C}$-algebras, $\mathbf{H}_q$ is isomorphic to $\mathbb{C}[K] \otimes_{\mathbb{C}} \mathbf{H}_q'$. The multiplication in the tensor product is the usual one*

$$(x \otimes h)(x' \otimes h') = xx' \otimes hh'.$$

Let L be an irreducible $\mathbf{H}_q$-module. Then every element x in $\mathbb{C}[K]$ acts on L by a scalar $x_L \in \mathbb{C}$ and the map $x \to x_L$ defines a homomorphism $\chi_L : \mathbb{C}[K] \to \mathbb{C}$. Denote L' the restriction to $\mathbf{H}_q'$ of L and by abuse notation we also denote χ_L an irreducible $\mathbb{C}[K]$-module corresponding to the homomorphism χ_L. By Assertion B we get

Assertion C. *The map $L \to (\chi_L, L')$ defines a bijection between the set $\mathrm{Irr}\mathbf{H}_q$ and the set $\mathrm{Irr}\mathbb{C}[K] \times \mathrm{Irr}\mathbf{H}_q'$. (For an algebra $\mathbf{A}$ we denote $\mathrm{Irr}\mathbf{A}$ the set of isomorphism classes of irreducible $\mathbf{A}$-modules.)*

Assume that G' is simply connected. Choose $q \in \mathbb{C}^*$ such that the Delinge-Langlands conjecture $(*)$ in the introduction is true for $\mathbf{H}_q$ and for $\mathbf{H}_q'$. (For example, when the order of q is not too small, see Theorem 6.6.) Let L be an irreducible $\mathbf{H}_q$-module corresponding to (s, N, ρ). It is no harm to assume that $s \in T$. Then L' is an irreducible $\mathbf{H}_q'$-module corresponding to (s', N, ρ) for some $s' \in T'$ with $s's^{-1} \in Z_G$. The element $s's^{-1}$ depends on the isomorphism $K \simeq \mathrm{Hom}(Z_G/Z_G \cap T', \mathbb{C}^*)$. I donot know how to determine the element $s's^{-1}$ canonically (it seems easy). A possible way is to consider the isomorphism $T \simeq (Z_G/Z_G \cap T') \times T'$.

Since G' is semisimple, there exist simple algebraic groups $G_1, ..., G_k$ such that G' is isomorphic to $G_1 \times \cdots \times G_k$. Let $W_1, ...W_k$ be the extended affine Weyl groups associated to $G_1, ..., G_k$, respectively, and let $\mathbf{H}_q^{(1)}, ..., \mathbf{H}_q^{(k)}$ be the corresponding affine Hecke algebras over $\mathbb{C}$ with parameter $q \in \mathbb{C}^*$. Obviously we have

Assertion D. *W' is isomorphic to $W_1 \times \cdots \times W_k$ and $\mathbf{H}_q'$ is isomorphic to $\mathbf{H}_q^{(1)} \otimes \cdots \otimes \mathbf{H}_q^{(k)}$. In particular, the map*

$$(L_1, ..., L_k) \to L_1 \otimes \cdots \otimes L_k$$

defines a bijection between the set $\mathrm{Irr}\mathbf{H}_q'$ and the set $\mathrm{Irr}\,\mathbf{H}_q^{(1)} \times \cdots \times \mathrm{Irr}\mathbf{H}_q^{(k)}$.

9.2. According to Assertion A-D in 9.1, it suffices to work with extended affine Weyl groups (and their Hecke algebras) associated to simple algebraic groups. In the rest part of the chapter G will be a simple algebraic group. Assume that $\varphi : G \to G'$ is an isogeny, i.e. φ is a surjective homomorphism of algebraic group and $\ker\varphi$ is finite. Let T be a maximal torus of G. Then $T' = \varphi(T)$ is a maximal torus of G'. We have

(a) The surjection $T \to T'$ induces an injective homomorphism

$$f : X' = \mathrm{Hom}(T', \mathbb{C}^*) \to X = \mathrm{Hom}(T, \mathbb{C}^*),$$

and induces an isomorphism from $W_0 = N_G(T)/T$ to $W_0' = N_{G'}(T')/T'$.

(b) If we identify W_0 with W_0', then the homomorphism f in (a) is W_0-equivariant.

Therefore we have

Assertion A. *The extended affine Weyl groups $W = W_0 \ltimes X$ and $W' = W_0' \ltimes X'$ are isogenous.*

Actually, let $R' \subset X'$ be the root system of W_0', then $f(R')$ is the root system of W_0. Now using (a) we see that the affine Weyl groups associated to W and to W' are isomorphic. That is, W and W' are isogenous.

Denote W_{aff} the affine Weyl group associated to W and to W'. Since G is simple, there exist two finite abelian groups Ω and Ω' such that we have

(c) $W \simeq \Omega \ltimes W_{\text{aff}}$ and $W' \simeq \Omega' \ltimes W_{\text{aff}}$. (Actually Ω and Ω' are isomorphic to the centers of G and of G' respectively.)

We shall regard W' as a subgroup group of W. Then Ω' is a subgroup of Ω. The following result will be needed.

(d) Let c' be a two-sided cell of W', then $\omega w \omega^{-1} \in c'$ for any w in c' and ω in Ω (see [L14, X1]). Moreover $\bigcup_{\omega \in \Omega} \omega c'$ is a two-sided cell of W. Further, if c is a two-sided cell of W, then $c \cap W'$ is a two-seded cell of W'.

Denote $\mathbf{H}_q$ (resp. $\mathbf{H}_q'$) the Hecke algebra over $\mathbb{C}$ of W (resp. W') with parameter $q \in \mathbb{C}^*$. Given an $\mathbf{H}_q$-module M, we denote $\text{Res}M$ the restriction to $\mathbf{H}_q'$ of M. For any $\mathbf{H}_q'$-module M', we denote $\text{Ind}M'$ the $\mathbf{H}_q$-module $\mathbf{H}_q \otimes_{\mathbf{H}_q'} M'$. Obviously, $\text{Ind}M'$ is spanned by $\omega \otimes m'$, $\omega \in \Omega, m \in M'$. For each $\omega \in \Omega$, let $\omega M'$ be the subspace of $\text{Ind}M'$ spanned by elements $\omega m' := \omega \otimes m'$, $m' \in M'$. Then $\omega M'$ is an $\mathbf{H}_q'$-submodule of $\text{Ind}M'$. In fact, $\omega^{-1} w \omega \in W'$ for any $\omega \in \Omega$ and $w \in W'$, so $T_w \omega m' = \omega T_{\omega^{-1} w \omega} m'$ is in ωM if $w \in W'$. Let $\{\omega_1, ..., \omega_k\}$ be a set of representatives of the (left) cosets of Ω' in Ω.

Proposition 9.3. *Keep the set up in 9.2. Let L be an irreducible $\mathbf{H}_q$-module and c the two-sided cell of W attached to L. Then*

(i). *$\text{Res}L$ is a completely reducible $\mathbf{H}_q'$-module.*

(ii). *Every irreducible submodule of $\text{Res}L$ is attached to the two-sided cell $c' := c \cap W'$ of W'.*

Proof. Let L' be an irreducible submodule of $\text{Res}L$. Then the map $\omega \otimes a \to \omega a$ defines a nonzero $\mathbf{H}_q$-homomorphism from $\text{Ind}L'$ to L. Since L is irreducible, the homomorphism must be surjective. Therefore it suffices to prove the proposition for $\text{Res}(\text{Ind}L')$. Let $\{\omega_1, ..., \omega_k\}$ be a set of representatives of the (left) cosets of Ω' in Ω, then
$$\text{Ind}L' = \omega_1 L' \oplus \cdots \oplus \omega_k L'.$$

Now L' is an irreducible $\mathbf{H}_q'$-module, so $\omega_i L'$ is an irreducible $\mathbf{H}_q'$-module for $i = 1, ..., k$. Hence (i) is true.

Since L' is a nonzero subset of L, we have $C_w L' \neq 0$ for some $w \in c$. By 9.2 (d), $w = \omega u$ for some $\omega \in \Omega$ and $u \in c'$. Since $C_w = C_\omega C_u$ and C_ω is invertible in $\mathbf{H}_q$, so $C_u L'$ is not zero. This implies (a). $C_{\omega_i u \omega_i^{-1} \omega_i} L'$ is not zero for $i = 1, ..., k$.

Let $v \in W'$ be such that $v \underset{LR}{\leq} u$ but $v \underset{LR}{\not\sim} u$. Then $C_{\omega^{-1} v \omega} L' = 0$ for every $\omega \in \Omega$. So we have

(b). $C_v \omega_i L' = \omega_i C_{\omega_i^{-1} v \omega_i} L' = 0$ for $i = 1, ..., k$.

Combining (a) and (b) we see that (ii) is true, and the proof is completed.

Proposition 9.4. *Keep the set up in 9.2. Let L' be an irreducible $\mathbf{H}'_q$-module. Assume that as $\mathbf{H}'_q$-modules L' is not isomorphic to $\omega L'$ for any $\omega \in \Omega - \Omega'$, then $\mathrm{Ind}L'$ is an irreducible $\mathbf{H}_q$-module.*

Proof. By the assumptions, as $\mathbf{H}'_q$-modules $\omega L'$ is not isomorphic to $\sigma L'$ if $\omega \sigma^{-1} \in \Omega - \Omega'$. Thus $\omega_1 L', ..., \omega_k L'$ are pairwise non-isomorphic $\mathbf{H}'_q$-modules. Let $a = \sum \omega_i a_i$ ($a_i \in L'$) be an element in $\mathrm{Ind}L'$. If a_i is not zero, then we can find a nonzero element b_i in L' such that $\omega_i b_i \in \mathbf{H}'_q a$. (Recall that $\{\omega_1, ..., \omega_k\}$ is a set of representatives of the (left) cosets of Ω' in Ω.) Therefore $\omega_i L'$ is an $\mathbf{H}'_q$-submodule of $\mathbf{H}'_q a$, and $\omega L'$ is contained in $\mathbf{H}_q a$ for each $\omega \in \Omega$. So $\mathbf{H}_q a = \mathrm{Ind}L'$. Hence $\mathrm{Ind}L'$ is an irreducible $\mathbf{H}_q$-module.

9.5. Keep the set up in 9.2. Assume that $\Omega \simeq \Lambda \times \Omega'$ for some subgroup Λ of Ω. Then $\mathbf{H}_q$ is isomorphic to the "twisted" tensor product $\mathbb{C}[\Lambda] \otimes_{\mathbb{C}} \mathbf{H}'_q$. (See 1.1 for the meaning of "twisted".) We have

Proposition 9.6. *Keep the set up in 9.5. Let L' be an irreducible $\mathbf{H}'_q$-module. Assume that as $\mathbf{H}'_q$-modules L' is isomorphic to $\omega L'$ for any $\omega \in \Lambda$, then*

(i). *$\mathrm{Ind}L'$ is a completely reducibe $\mathbf{H}_q$-module, and each irreducible submodule of $\mathrm{Ind}L'$ appears in a composition factors series of $\mathrm{Ind}L'$ with multiplicity 1.*

(ii). *If L' is attached to a two-sided cell c' of W', then every irreducible submodule of $\mathrm{Ind}L'$ is attached to the two-sided cell $c := \bigcup_{\omega \in \Lambda} \omega c'$ of W.*

Proof. Λ is either a cyclic group or isomorphic to the group $\mathbb{Z}/(2) \times \mathbb{Z}/(2)$.

Assume that Λ is a cyclic group and let ω be a generator of Λ. Let $f_{0,1} : L' \simeq \omega L'$ be an isomorphism of $\mathbf{H}'_q$-module. For every $a \in L'$, denote the image $f(a)$ by ωa_ω. Write a_{ω^2} for $(a_\omega)_\omega$, a_{ω^3} for $(a_{\omega^2})_\omega$, and so on. For any integers i, j, then the map

$$f_{i,j} : \omega^i L' \to \omega^j L', \quad \omega^i a_{\omega^i} \to \omega^j a_{\omega^j}$$

is an isomorphism of $\mathbf{H}'_q$-module. We normalize the isomorphisms so that $a_{\omega^m} = a$ for all a in L', where $m = |\Lambda|$. This is possible since L' is an irreducible $\mathbf{H}'_q$-module. Then we have $f_{j,k} f_{i,j} = f_{i,k}$.

Let $\xi \in \mathbb{C}^*$ be an m-th root of 1 and let L_ξ be the subspace of $\mathrm{Ind}L'$ spanned by

$$a(\xi) := a + \xi \omega a_\omega + \cdots + \xi^{m-1} \omega^{m-1} a_{\omega^{m-1}}, \quad a \in L'.$$

Then L_ξ is an $\mathbf{H}'_q$-submodule of $\mathrm{Ind}L'$. Moreover $\omega a(\xi) = \xi^{m-1} a_{\omega^{m-1}}(\xi)$. Hence L_ξ is an $\mathbf{H}_q$-submodule of $\mathrm{Ind}L'$. As an $\mathbf{H}'_q$-module, L_ξ is isomorphic to L', so L_ξ is an irreducible $\mathbf{H}_q$-module.

Let $\eta \in \mathbb{C}^*$ be an m-th root of 1 and suppose that $\psi : L_\xi \to L_\eta$ is an isomorphism of $\mathbf{H}_q$-module. Let a, b be two nonzero elements in L' such that $\psi(a(\xi)) = b(\eta)$. Then we have $f(\omega a(\xi)) = \omega b(\eta)$. Since L' is an irreducible $\mathbf{H}'_q$-module, we can find some h in $\mathbf{H}'_q$ such that $ha = b$. Thus we have $ha_{\omega^i} = b_{\omega^i}$ for $i = 0, 1, ..., m-1$. So

$$f(\omega a(\xi)) = \xi^{m-1} f(a_{\omega^{m-1}}(\xi)) = \xi^{m-1} b_{\omega^{m-1}}(\eta) = \omega b(\eta) = \eta^{m-1} b_{\omega^{m-1}}(\eta).$$

Therefore $\xi = \eta$.

We have $\mathrm{Ind} L' = \bigoplus_{\substack{\xi \in \mathbb{C}^* \\ \xi^m = 1}} L_\xi$, so $\mathrm{Ind} L'$ is a completely reducible $\mathbf{H}_q$-module and each irreducible submodule of $\mathrm{Ind} L'$ appears in a composition factors series of $\mathrm{Ind} L'$ with multiplicity 1.

Note that $C_\sigma C_w = C_{\sigma w}$ for any $\sigma \in \Lambda, w \in W$, and C_σ is invertible in $\mathbf{H}_q$. Since L_ξ is isomorphic to L' as an $\mathbf{H}'_q$-module, we know that the two-sided cell of W attached to L_ξ is c for any m-th root ξ of 1. The proposition is proved provided that Λ is cyclic.

Similar argument is valid when Λ is isomorphic to $\mathbb{Z}/(2) \times \mathbb{Z}/(2)$.

9.7. Denote $\mathbf{S}_q$ the Hecke algebra (over $\mathbb{C}$) of W_{aff} with parameter q. Then $\mathbf{H}_q$ (resp. $\mathbf{H}'_q$) is isomorphic to the "twisted" tensot product $\mathbb{C}[\Omega] \otimes_{\mathbb{C}} \mathbf{S}_q$ (resp. $\mathbb{C}[\Omega'] \otimes_{\mathbb{C}} \mathbf{S}_q$). Let L' be an irreducible $\mathbf{H}'_q$-module. Denote $\mathrm{res} L'$ the restriction to $\mathbf{S}_q$ of L'. According to Proposition 9.3, $\mathrm{res} L'$ is a completely reducible $\mathbf{S}_q$-module. Let E be an irreducible submodule of $\mathrm{res} L'$ and set

$$\Lambda_E = \{\omega \in \Omega \mid E \text{ and } \omega E \text{ are isomorphic } \mathbf{S}_q\text{-modules}\},$$

$$\Lambda'_E = \{\omega \in \Omega' \mid E \text{ and } \omega E \text{ are isomorphic } \mathbf{S}_q\text{-modules}\}.$$

Denote the subalgebra $\mathbb{C}[\Lambda_E] \otimes_{\mathbb{C}} \mathbf{S}_q$ of $\mathbf{H}_q$ by $\mathbf{T}$ and denote the subalgebra $\mathbb{C}[\Lambda'_E] \otimes_{\mathbb{C}} \mathbf{S}_q$ of $\mathbf{H}'_q$ by $\mathbf{T}'$. According to Proposition 9.6 and its proof, we have

(a) $F' := \mathbf{T}' \otimes_{\mathbf{S}_q} E$ is a completely reducible $\mathbf{T}'$-module, and irreducible submodules of $\mathbf{T}' \otimes_{\mathbf{S}_q} E$ are pairwise non-isomorphic. Moreover, as an $\mathbf{S}_q$-module, each irreducible submodule of $\mathbf{T}' \otimes_{\mathbf{S}_q} E$ is isomorphic to E.

(b) $F := \mathbf{T} \otimes_{\mathbf{S}_q} E$ is a completely reducible $\mathbf{T}$-module, and irreducible submodules of $\mathbf{T} \otimes_{\mathbf{S}_q} E$ are pairwise non-isomorphic. Moreover, as an $\mathbf{S}_q$-module, each irreducible submodule of $\mathbf{T} \otimes_{\mathbf{S}_q} E$ is isomorphic to E.

Using (a), (b) and Prosposition 9.4 we get

(c) $D' := \mathbf{H}'_q \otimes_{\mathbf{S}_q} E(= \mathbf{H}'_q \otimes_{\mathbf{T}'} F')$ is a completely reducible $\mathbf{H}'_q$-module and $l(D') = |\Lambda'_E|$. (We use $l(M)$ for the length of a composition series of a module M.)

(d) $D := \mathbf{H}_q \otimes_{\mathbf{S}_q} E(= \mathbf{H}_q \otimes_{\mathbf{T}} F)$ is a completely reducible $\mathbf{H}_q$-module and $l(D) = |\Lambda_E|$.

Since L' is a submodule of D' and $D = \mathbf{H}_q \otimes_{\mathbf{H}'_q} D'$, by (d) we obtain

(e) $\operatorname{Ind}L' = \mathbf{H}_q \otimes_{\mathbf{H}'_q} L'$ is a completely reducible $\mathbf{H}_q$-module and $l(\operatorname{Ind}L') = |\Lambda_E|/|\Lambda'_E|$.

It is not difficult to describe the simple consitituents of $\operatorname{Ind}L'$.

9.8. There are several interesting consequences. Assume that G is simply connected and G' is adjoint. Suppose that $\sum_{w \in W_0} q^{l(w)}$ is not zero. Let E be an irreducible $\mathbf{H}'_q$-module attached to the lowest two-sided cell c'_0 of W'. Let s' be the semisimple conjugacy class of G' such that the center of $\mathbf{H}'_q$ acts on E through s'. Then $\Lambda_E = \Omega$ if and only if $\varphi^{-1}(s')$ contains $|\Omega|$ semisimple conjugacy classes of G and $\Lambda_E = \{e\}$ if and only if $\varphi^{-1}(s')$ is a semisimple conjugacy class of G.

9.9. Similarly we may discuss affine Hecke algebras with two parameters.

10. Quotient Algebras

Given $q \in \mathbb{C}^*$, the problem of classifying simple $\mathbf{H}_q$-modules is equivalent to the problem of classifying simple $\mathbf{H}_{s,q}$-modules for all semisimple elements s in G. Note that the standard module $M_{s,N,q,\rho}$ is actually an $\mathbf{H}_{s,q}$-module. Thus it is also interesting to study the algebras $\mathbf{H}_{s,q}$. Ginsburg gave a nice geometric realization for the algebras $\mathbf{H}_{s,q}$. In chapter 6 we have showed that the the number of the isomorphism classes of these algebras $\mathbf{H}_{s,q}$ is finite. It would be interesting to classify the algebras $\mathbf{H}_{s,q}$, $(s,q) \in G \times \mathbb{C}^*$ semisimple. In this chapter we give some discussions to the algebras. In the same way we discuss the algebra $\tilde{\mathbf{H}}_{a,b}$ and its quotient algebras $\tilde{\mathbf{H}}_{s,a,b}$.

Notations are as in chapter 5. We assume that G is simply connected and simple.

10.1. Completions.
For arbitrary $x, y \in X^+$, in $\mathbf{H}_q$ we have

$$U_x U_y = U_{x+y} + \sum_{z < x+y} a_z U_y, \qquad a_z \in \mathbb{C}.$$

This implies that

$$(10.1.1) \qquad \bigcap_{k=1}^{\infty} \mathbf{I}_{s,q}^k = \emptyset.$$

where $s \in G$ is a semisimple element and $\mathbf{I}_{s,q}$ is the two-sided ideal of $\mathbf{H}_q$ generated by $U_x - tr(s, V(x))$, $x \in X^+$.

We have the following natural inverse limit system

$$\cdots \to \mathbf{H}_q / \mathbf{I}_{s,q}^k \to \cdots \to \mathbf{H}_q / \mathbf{I}_{s,q}^2 \to \mathbf{H}_q / \mathbf{I}_{s,q}.$$

By (10.1.1) we see that the completion $\hat{\mathbf{H}}_{s,q}$ of $\mathbf{H}_q$ with respect to the two-sided ideal $\mathbf{I}_{s,q}$ is just the inverse limit $\varprojlim \mathbf{H}_q / \mathbf{I}_{s,q}^k$. Let $\mathrm{Mod}(\mathbf{H}_q)$ be the category of finite dimensional $\mathbf{H}_q$-modules (over $\mathbb{C}$) and let $\mathrm{Mod}(\hat{\mathbf{H}}_{s,q})$ be category of finite dimensional $\hat{\mathbf{H}}_{s,q}$-modules (over $\mathbb{C}$).

We have the following

10.2. Theorem.
The category $\mathrm{Mod}(\mathbf{H}_q)$ *of finite dimensional* $\mathbf{H}_q$*-modules (over* $\mathbb{C}$*) is the direct sum of the categories* $\mathrm{Mod}(\hat{\mathbf{H}}_{s,q})$*, where the direct sum is over the set* $\mathcal{S}$ *of representatives of semisimple conjugacy classes of* G.

Proof. First, each finite dimensional $\hat{\mathbf{H}}_{s,q}$-module has a natural $\mathbf{H}_q$-module structure.

Let M be a finite dimensional $\mathbf{H}_q$-module. For $s \in \mathcal{S}$ semisimple, we write M_s for the set

$$\{m \in M \mid (U_x - tr(s, V(x)))^k m = 0 \text{ for all } x \in X^+ \text{ and for some integer } k > 0\}.$$

The space M_s is an $\mathbf{H}_q$-module. We have

$$M = \bigoplus_{s \in S} M_s.$$

Moreover, if $s, t \in S$ are different, we have

$$\mathrm{Hom}_{\mathbf{H}_q}(M_s, M_t) = 0.$$

Since M is of finite dimension, M_s is actually an $\hat{\mathbf{H}}_{s,q}$-module. The theorem is proved.

Note that $\mathbf{H}_{s,q}$ is also a quotient algebra of $\hat{\mathbf{H}}_{s,q}$.

10.3. Let E be a simple $\mathbf{H}_q$-module, then $E \in Y_{s,q}$ for some semisimple element $s \in S$. We regard E as an $\mathbf{H}_{s,q}$-module in a natural way. Then obviously the set of isomorphism classes of simple $\mathbf{H}_{s,q}$-modules is $Y_{s,q}$. By a theorem of Pittie (see [St2]) we know that $\dim \mathbf{H}_{s,q} = |W_0|^2$. Thus $\sum_{E \in Y_{s,q}} (\dim E)^2 \leq |W_0|^2$. The equality holds if and only if $\mathbf{H}_{s,q}$ is semisimple. In particular, we see that any simple $\mathbf{H}_q$-module has dimension $\leq |W_0|$ and the equality holds if and only if $\mathbf{H}_{s,q}$ is simple. According to Ginzburg [G3] we know that the algebra $\mathbf{H}_{s,q}$ is either simple or non-semisimple. Thus the equality $\sum_{E \in Y_{s,q}} (\dim E)^2 = |W_0|^2$ holds if and only if $\mathbf{H}_{s,q}$ is simple.

When $\mathbf{H}_{s,q}$ is simple, we have $\#Y_{s,q} = 1$. Moreover the natural map $\mathbf{H}_q^{\leq c_0} \to \mathbf{H}_{s,q}$ is a surjective map. Let $E \in Y_{s,q}$, then we have $\dim E = |W_0|$ and $c_E = c_0$. Let $\mathbf{H}_{q,W_0}$ be the subalgebra of $\mathbf{H}_{s,q}$ generated by the images of T_w, $w \in W_0$. The algebra $\mathbf{H}_{q,W_0}$ is isomorphic to the Hecke algebra of W_0 over $\mathbb{C}$ with parameter q. It is easy to see that as an $\mathbf{H}_{q,W_0}$-module, E is isomorphic to the left regular module of $\mathbf{H}_{q,W_0}$. In general $\#Y_{s,q} = 1$ does not imply that $\mathbf{H}_{s,q}$ is simple. We summarize the discussions as follows.

10.4. Proposition. *Let E be a simple $\mathbf{H}_q$-module and s an element in S such that the center of $\mathbf{H}_q$ acts on E through s. Then the following conditions are equivalent.*

(i). *$\dim E = |W_0|$.*

(ii). *$\mathbf{H}_{s,q}$ is a simple algebra.*

(iii). *As an $\mathbf{H}_{q,W_0}$-module, E is isomorphic to the (left) regular module of $\mathbf{H}_{q,W_0}$.*

(iv). *$\mathcal{N}_{s,q} = \{0\}$.*

(v). *$\mathbf{H}_q^{\leq c_0} \to \mathbf{H}_{s,q}$ is a surjective map.*

10.5. Conjecture. (i). *Let $(s,q), (t,r) \in G \times \mathbb{C}^*$ be semisimple, then $\hat{\mathbf{H}}_{s,q}$ is isomorphic to $\hat{\mathbf{H}}_{t,r}$ if $\mathcal{N}_{s,q} = \mathcal{N}_{t,r}$.*

(ii). *Consider the homomorphism* $\phi_{q,c_0} : \mathbf{H}_q \to \mathbf{J}_{c_0}$. *It induces a homomorphism* $\phi_{s,q,c_0} : \mathbf{H}_{s,q} \to \mathbf{J}_{c_0}/\phi_{q,c_0}(\mathbf{I}_{s,q})$. *The ideal* ker$\phi_{s,q,c_0}$ *of* $\mathbf{H}_{s,q}$ *should be nilpotent. Note that* $\mathbf{J}_{c_0}/\phi_{q,c_0}(\mathbf{I}_{s,q})$ *is a simple* $\mathbb{C}$*-algebra of dimension* $|W_0|^2$ *(see [X2]).*

10.6. One may consider the algebra $\tilde{\mathbf{H}}_{a,b}$, $(a,b) \in \mathbb{C}^* \times \mathbb{C}^*$ in a similar way. Given a semisimple element s in G, let $\tilde{\mathbf{I}}_{s,a,b}$ be the two-sided ideal of $\tilde{\mathbf{H}}_{a,b}$ generated by $U_x - tr(s, V(x))$, $x \in X^+$. For the same reason of (10.1.1) we know that the completion $\hat{\tilde{\mathbf{H}}}_{s,a,b}$ of $\tilde{\mathbf{H}}_{a,b}$ with respect to the two-sided ideal $\tilde{\mathbf{I}}_{s,a,b}$ is just the inverse limit $\varprojlim \tilde{\mathbf{H}}_{a,b}/\tilde{\mathbf{I}}_{s,a,b}^k$. Let Mod($\tilde{\mathbf{H}}_{a,b}$) be the category of finite dimensional $\tilde{\mathbf{H}}_{a,b}$-modules (over $\mathbb{C}$) and let Mod($\hat{\tilde{\mathbf{H}}}_{s,a,b}$) be category of finite dimensional $\hat{\tilde{\mathbf{H}}}_{s,a,b}$-modules (over $\mathbb{C}$). Similar to 10.2 we have the following

(a). *The category* Mod($\tilde{\mathbf{H}}_{a,b}$) *of finite dimensional* $\tilde{\mathbf{H}}_{a,b}$*-modules (over* $\mathbb{C}$*) is the direct sum of the categories* Mod($\hat{\tilde{\mathbf{H}}}_{s,a,b}$)*, where the direct sum is over the set* $\mathcal{S}$ *of representatives of semisimple conjugacy classes of* G.

Note that $\tilde{\mathbf{H}}_{s,a,b} = \tilde{\mathbf{H}}_{a,b}/\tilde{\mathbf{I}}_{s,a,b}$ is a quotient algebra of $\hat{\tilde{\mathbf{H}}}_{s,a,b}$.

Let E be a simple $\tilde{\mathbf{H}}_{a,b}$-module, then $E \in Y_{s,a,b}$ for some semisimple element $s \in \mathcal{S}$ ($Y_{s,a,b}$ is the set of isomorphism classes of simple $\tilde{\mathbf{H}}_{a,b}$-modules on which all elements in $\tilde{\mathbf{I}}_{s,a,b}$ act by scalar 0). We may regard E as an $\tilde{\mathbf{H}}_{s,a,b}$-module in a natural way. Then obviously the set of isomorphism classes of simple $\tilde{\mathbf{H}}_{s,a,b}$-modules is $Y_{s,a,b}$. According to a theorem of Pittie (see [St2]) we know that dim$\tilde{\mathbf{H}}_{s,a,b} = |W_0|^2$. Thus $\sum_{E \in Y_{s,a,b}} (\dim E)^2 \leq |W_0|^2$. The equality holds if and only if $\mathbf{H}_{s,q}$ is semisimple. In particular, we see that any simple $\mathbf{H}_q$-module has dimension $\leq |W_0|$ and the equality holds if and only if $\tilde{\mathbf{H}}_{s,a,b}$ is simple.

Let $s \in T$, then (see [Ka1])

(b). $\tilde{\mathbf{H}}_{s,a,b}$ is simple if and only if $\mathcal{N}_{s,a,b} = \{0\}$ (see (5.19.7) for the notation).

10.7. Conjecture. *We keep the notations in 10.6. Let* (s,a,b) *and* (t,c,d) *be two elements in* $T \times \mathbb{C}^* \times \mathbb{C}^*$. *Assume that* $\mathcal{N}_{s,a,b} = \mathcal{N}_{t,c,d}$, *then*

(i). *The* $\mathbb{C}$*-algebras* $\tilde{\mathbf{H}}_{s,a,b}$ *and* $\tilde{\mathbf{H}}_{t,c,d}$ *are isomorphic.*

(ii). *The* $\mathbb{C}$*-algebras* $\hat{\tilde{\mathbf{H}}}_{s,a,b}$ *and* $\hat{\tilde{\mathbf{H}}}_{t,c,d}$ *are isomorphic.*

11. The Based Rings of Cells in Affine Weyl Groups
of Type $\widetilde{G}_2$, $\widetilde{B}_2$, $\widetilde{A}_2$

The works [L12-14] and [X3] show that the based rings of two-sided cells in affine Weyl groups are interesting in understanding the classifications of simple modules of the corresponding Hecke algebras $\mathbf{H}_q$ ($q \in \mathbb{C}^*$) even if q is a root of 1.

In this chapter we determine the based rings of cells in affine Weyl groups of type $\tilde{G}_2$, $\tilde{B}_2$ (see 11.2), which confirm the conjecture in [L14] (see also 5.14). Then we apply the results to classify the simple modules of the corresponding Hecke algebra $\mathbf{H}_q$ ($q \in \mathbb{C}^*$).

The explicit descriptions of the based rings enable us to understand the structure of the standard modules in a concrete way and provide a way to compute the dimensions of simple modules of $\mathbf{H}_q$. As an example we work out the case of type $\tilde{A}_2$ (see 11.7). An immediate consequence is that $\mathbf{H}_q \not\simeq \mathbf{H}_1 = \mathbb{C}[W]$ whenever $q \neq 1$, here $\mathbf{H}_q$ is an affine Hecke algebra of type $\tilde{A}_2$. This result leads to several questions (see 11.7). This chapter is based on the preprint [X4], only section 11.7 is added. We refer to chapters 2, 5 for notations.

The based ring J_c

11.1. We refer to 5.14 for Lusztig's conjecture concerned with based rings of cells in affine Weyl groups. The conjecture is a guideline to determine the structure of the based rings. Except special indications, until section 11.5, G is always a simple, simply connected algebraic group over $\mathbb{C}$ of type G_2 or B_2, and W is an extended affine Weyl group of type $\tilde{G}_2$ or $\tilde{B}_2$. Denote $S = \{r_0, r_1, r_2\}$ the set of simple reflections in W. The cells in W have been described in [L11] explicitly.

For type $\tilde{G}_2$, we have $W = W' = W_0 \ltimes P$. Assume that $(r_0 r_1)^3 = (r_1 r_2)^6 = (r_0 r_2)^2 = e$. The affine Weyl group W has five two-sided cells:

$$c_e = \{w \in W \mid a(w) = 0\} = \{e\},$$

$$c_1 = \{w \in W \mid a(w) = 1\},$$

$$c_2 = \{w \in W \mid a(w) = 2\},$$

$$c_3 = \{w \in W \mid a(w) = 3\},$$

$$c_0 = \{w \in W \mid a(w) = 6\}.$$

(See 1.12 for the definition of the function $a : W \to \mathbb{N}$.)

For type $\tilde{B}_2$, we have $W = \Omega \ltimes W'$ and $\Omega = \{e, \omega\}$. Assume that $\omega r_0 = r_2 \omega$, $\omega r_1 = r_1 \omega$ and $\omega r_2 = \omega r_0$. There are four two-sided cells in W:

$$c_e = \{w \in W \mid a(w) = 0\} = \{e, \omega\} = \Omega,$$

$$c_1 = \{w \in W \mid a(w) = 1\},$$

$$c_2 = \{w \in W \mid a(w) = 2\},$$

$$c_0 = \{w \in W \mid a(w) = 4\}.$$

One main result of this chapter is the following.

102

Theorem 11.2. *We keep the set up in 11.1. For each two-sided cell c of W, there exists a finite F_c-set Y (see 5.4 for the notion) and a bijection*

$$\pi: c \overset{\sim}{\to} \text{ set of irreducible } F_c\text{-v.b. on } Y \times Y \text{ (up to isomorphisms)}$$

with the following properties:

(i). The $\mathbb{C}$-linear map $\mathbf{J}_c \to \mathbf{K}_{F_c}(Y \times Y)$, $t_w \to \pi(w)$ is an algebra isomorphism (preserving the unit element).

(ii). $\pi(w^{-1}) = \widetilde{\pi(w)}$ for any $w \in c$.

When Γ is a left cell in c_1, the ring $\mathbf{J}_{\Gamma \cap \Gamma^{-1}}$ has been described in [L15], here $\mathbf{J}_{\Gamma \cap \Gamma^{-1}}$ is the $\mathbb{C}$-subspace of $\mathbf{J}_{c_1}$ spanned by elements t_w, $w \in \Gamma \cap \Gamma^{-1}$.

Proof. We prove the theorem case by case. Before our proof we make a convention. We often write $i_1 i_2 \ldots i_n$ and $C_{r_{i_1} r_{i_2} \ldots r_{i_n}}$ instead of w and C_w when $r_{i_1} r_{i_2} \ldots r_{i_n}$ is a reduced expression of $w \in W$.

(A). When $c = c_e$, the reductive group F_c is the center of G. We take Y to be a one point set and let F_c act on Y trivially. The theorem then is trivial. When c is the lowest two-sided cell c_0 of W, then $F_c = G$ and the theorem is proved in [X2]. Thus we only need to verify the theorem for the two-sided cells c of W with $c \neq c_e, c_0$. In (B-D), we assume that G is of type G_2, then the affine Weyl group W is of type $\tilde{G}_2$.

(B). Case $c = c_1$. We have $c = \{2, 212, 21212, 21, 2121, 210, 21210, 1, 121, 12121, 12, 1212, 10, 1210, 121210, 0, 01210, 0121210, 01, 0121, 012121, 012, 01212\}$ (recall the convention at the beginning of the proof), and $F_c = \mathfrak{S}_3$, the symmetric group of three letters. Let $Y = \{1, 2, 3, 4, 5\}$ be the F-set such that as F-sets we have $\{1\} \simeq \{2\} \simeq \mathfrak{S}_3/\mathfrak{S}_3$ and $\{3, 4, 5\} \simeq \mathfrak{S}_3/\mathfrak{S}_2$. We assume that $\mathfrak{S}_2$ leaves stable on 3.

For $\mathfrak{S}_3$ we have three simple representations: the unit representation 1, the sign representation ε, and the unique simple representation σ of degree 2. The notations 1, ε also stand for their restrictions to $\mathfrak{S}_2$ respectively. One may verify that the following bijection (recall the convention in 5.4)

$$\pi: c \overset{\sim}{\to} \text{ set of irreducible } F_c\text{-v.b. on } Y \times Y \text{ (up to isomorphisms)},$$

$$
\begin{array}{lll}
0 \to 1_{(1,1)}, & 01210 \to \sigma_{(1,1)}, & 0121210 \to \varepsilon_{(1,1)}, \\
1 \to 1_{(2,2)}, & 121 \to \sigma_{(2,2)}, & 12121 \to \varepsilon_{(2,2)}, \\
01 \to 1_{(1,2)}, & 0121 \to \sigma_{(1,2)}, & 012121 \to \varepsilon_{(1,2)}, \\
10 \to 1_{(2,1)}, & 1210 \to \sigma_{(2,1)}, & 121210 \to \varepsilon_{(2,1)}, \\
2 \to 1_{(3,3)}, & 212 \to 1_{(3,4)}, & 21212 \to \varepsilon_{(3,3)}, \\
012 \to 1_{(1,3)}, & 01212 \to \varepsilon_{(1,3)}, & \\
12 \to 1_{(2,3)}, & 1212 \to \varepsilon_{(2,3)}, & \\
210 \to 1_{(3,1)}, & 21210 \to \varepsilon_{(3,1)}, & \\
21 \to 1_{(3,2)}, & 2121 \to \varepsilon_{(3,2)}, & \cdot
\end{array}
$$

103

is just what we need.

Let $Y' = \{z_i \mid 1 \le i \le 7\}$ be the F_c-set such that as F_c-sets we have $\{z_1\} \simeq \mathfrak{S}_3/\mathfrak{S}_3$ and $\{z_2, z_3, z_4\} \simeq \{z_5, z_6, z_7\} \simeq \mathfrak{S}_3/\mathfrak{S}_2$. One may check that there exists a bijection between c and the set of isomorphism classes of irreducible F-v.b. on $Y' \times Y'$ with the properties (i) and (ii) in Theorem 11.2.

(C). Case $c = c_2$. We have $F_c = SL_2(\mathbb{C})$ and

$$c = \{w(i,j,k) \mid 1 \le i, j \le 6, \ k \in \mathbb{N}\},$$

where $w(i,j,k) = w_i r_0 r_2 (r_1 r_2 r_1 r_2 r_0)^k w_j^{-1}$, $w_1 = e$, $w_2 = r_1$, $w_3 = r_2 r_1$, $w_4 = r_1 r_2 r_1$, $w_5 = r_2 r_1 r_2 r_1$, and $w_6 = r_0 r_1 r_2 r_1$. We write $w(k)$ for $w(1,1,k)$. Then

(a) $\mathcal{D} \cap c = \{w(i,i,0) \mid 1 \le i \le 6\} = \{w_i r_0 r_2 w_i^{-1} \mid 1 \le i \le 6\}$. (See 2.6 for the definition of $\mathcal{D}$.)

(b) $w(i,j,k) \underset{L}{\sim} w(m,n,k')$ if and only if $j = n$,

$w(i,j,k) \underset{R}{\sim} w(m,n,k')$ if and only if $i = m$.

Let $Y = \{1,2,3,4,5,6\}$ and let F_c act on Y trivially. Then the map

$$\pi: w(i,j,k) \to V(k)_{(i,j)}$$

defines a bijection between the two-sided cell c and the set of isomorphism classes of irreducible F_c-v.b. on $Y \times Y$, where $V(k)$ is the irreducible representation of F_c with highest weight k. We claim that the bijection is what we need. In fact, 11.2(ii) is obvious. The map π gives rise to a $\mathbb{C}$-linear map $\pi: \mathbf{J}_c \to \mathbf{K}_{F_c}(Y \times Y)$ which preserves the unit element. To complete the proof we need to check the following equality.

(c) $\qquad t_{w(i,j,k)} t_{w(m,n,k')} = \pi(w(i,j,k))\pi(w(m,n,k'))$

When $j \ne m$, using (b), 2.7(a) and the definition of π, we see that (c) is true since both sides in (c) are 0. Now we assume that $j = m$, then (c) is equivalent to the following.

(d) $\qquad t_{w(i,j,k)} t_{w(j,n,k')} = \displaystyle\sum_{\substack{k'' \in \mathbb{N} \\ |k-k'| \le k'' \le k+k'}} t_{w(i,n,k'')}.$

We claim that (d) deduces from the following two assertions.

(e) $\gamma_{w(i,j,k), w(j,n,k'), w(i,n,k'')} = \gamma_{w(k), w(k'), w(k'')}$.

(f) $t_{w(1)} t_{w(k')} = t_{w(k'+1)} + t_{w(k'-1)}$. (We assume that $t_{w(-1)} = 0$.)

In fact, according to (a) and 2.7(b), we have $t_{w(0)} t_{w(k')} = t_{w(k')} = t_{w(k')} t_{w(0)}$. Using induction on k and using (f), we get

(g) $\qquad t_{w(k)} t_{w(k')} = \displaystyle\sum_{\substack{k'' \in \mathbb{N} \\ |k-k'| \le k'' \le k+k'}} t_{w(k'')}$

104

Combine (e), (g) and 2.7(a) and we see that (d) holds.

Now we return to (e) and (f). First we prove (e). Consider the algebra $H_{\geq c} = H/H^{<c}$ (see 2.6(h)). For simplicity we also denote the image in $H_{\geq c}$ of C_w by the same notation. Then in $H_{\geq c}$ there exist h_i and h'_j such that

(h) $$C_{w(i,j,k)} = h_i C_{w(k)} h'_j.$$

By (h) and the definition of γ we obtain

(i) $$\gamma_{w(i,j,k),w(j,n,k'),w(i,n,k'')} = \gamma_{w(1,j,k),w(j,1,k'),w(k'')}.$$

Using [X3, 2.6], we see that there exists h in $H_{\geq c}$ such that

(j) $\qquad C_{w(k)} = h C_{02}$ (note the convention at the beginning of the proof).

By (h) and (j) we know that

(k) $$C_{w(k)} h'_j = C_{w(1,j,k)} = h C_{w(1,j,0)}.$$

From (b) and 2.6(f-g), we have the following identity in $H_{\geq c}$

(ℓ) $$C_{w(1,j,0)} C_{w(j,1,k')} = \sum_{m \in \mathbb{N}} a_m C_{w(m)}, \ a_m \in A.$$

According to (a), (b) and 2.7(a), we get the following fact.

(m) When $m \neq k'$, we have $q a_m \in q^{\frac{1}{2}} \mathbb{C}[q^{\frac{1}{2}}]$ and $q^{\frac{1}{2}} a_{k'} \in 1 + q^{\frac{1}{2}} \mathbb{C}[q^{\frac{1}{2}}]$.

By (k) and (ℓ) we get

(n) $C_{w(1,j,k)} C_{w(j,1,k')} = \sum\limits_{m \in \mathbb{N}} a_m h C_{w(m)}$

It is easy to see (1.8(a))

(o) $$C_{02} C_{w(m)} = C_{w(m)} C_{02} = (q^{\frac{1}{2}} + q^{-\frac{1}{2}})^2 C_{w(m)}.$$

From (o), (n), (j), (b) and 2.6(f-g), we obtain

(p) $\qquad C_{w(1,j,k)} C_{w(j,1,k')} = (q^{\frac{1}{2}} + q^{-\frac{1}{2}})^{-2} \sum\limits_{m \in \mathbb{N}} a_m C_{w(k)} C_{w(m)}$

$$= (q^{\frac{1}{2}} + q^{-\frac{1}{2}})^{-2} \sum_{m,k'' \in \mathbb{N}} a_m h_{w(k),w(m),w(k'')} C_{w(k'')}$$

Using (o) and (j) again, we see that $h_{w(k),w(m),w(k'')} = (q^{\frac{1}{2}} + q^{-\frac{1}{2}})^2 \cdot a_{k,m,k''}$ for some $a_{k,m,k''}$ in A. Since $h_{w(k),w(m),w(k'')}$ is polynomial in $q^{\frac{1}{2}} + q^{-\frac{1}{2}}$, by the definition of c its degree is ≤ 2, thus $a_{k,m,k''} \in \mathbb{Z}$. (In fact, $a_{k,m,k''} \in \mathbb{N}$ by the positivity of $h_{w(k),w(m),w(k'')}$ (cf. [L11]). Therefore we have

(q)
$$C_{w(1,j,k)}C_{w(j,1,k')} = \sum_{m,k'' \in \mathbb{N}} a_m a_{k,m,k''} C_{w(k'')}$$

Combining (m) and (q) we get

(r)
$$\gamma_{w(1,j,k),w(j,1,k'),w(k'')} = \gamma_{w(k),w(k'),w(k'')}.$$

Then (e) follows from (i) and (r).

Now we prove (f). In $H_{\geq c}$ we have

(s)
$$C_{w(1)} = (C_0 C_2 C_1 C_2 C_1 - 2(\mathbf{q}^{\frac{1}{2}} + \mathbf{q}^{-\frac{1}{2}}))C_{02} = C_{0212120}$$

Using (o) and (s), we get

(t)
$$C_{w(1)}C_{w(k')} = (\mathbf{q}^{\frac{1}{2}} + \mathbf{q}^{-\frac{1}{2}})^2 (C_1 C_2 C_1 C_2 C_1 - 2(\mathbf{q}^{\frac{1}{2}} + \mathbf{q}^{-\frac{1}{2}}))C_{w(k')}.$$

From (t) and 1.8(a), it is not difficult to see that

(u)
$$C_{w(1)}C_{w(k')} = (\mathbf{q}^{\frac{1}{2}} + \mathbf{q}^{-\frac{1}{2}})^2 C_0 C_{2121 w(k')} = (\mathbf{q}^{\frac{1}{2}} + \mathbf{q}^{-\frac{1}{2}})^2 C_0 C_{w(5,1,k')}.$$

Using 1.8(a) again, we get the following identity in $H_{\geq c}$

(v)
$$C_0 C_{w(5,1,k')} = C_{w(k'+1)} + \sum_{w(m) \prec w(5,1,k')} \mu(w(m), w(5,1,k')) C_{w(m)}.$$

By [L11, 10.4.3] we see that $\mu(w(m), w(5,1,k')) = \mu(w(5,1,m), w(k'))$. According to 1.8 (a), we see that $\mu(w(5,1,m), w(k')) \neq 0$ if and only if $r_0 w(5,1,m) = w(k')$, i.e. $m = k' - 1$; moreover, $\mu(w(5,1,k'-1), w(k') = 1$. Hence

$$C_{w(1)}C_{w(k')} = (\mathbf{q}^{\frac{1}{2}} + \mathbf{q}^{-\frac{1}{2}})^2 (C_{w(k'+1)} + C_{w(k'-1)})$$

and (f) follows.

(D). Case $c = c_3$. We have $F_c = SL_2(\mathbb{C})$ and

$$c = \{u(i,j,k) \mid 1 \leq i, j \leq 6, \ k \in \mathbb{N}\},$$

where $u(i,j,k) = u_i r_0 r_1 r_0 (r_2 r_1 r_0)^k u_j^{-1}$, $u_1 = e$, $u_2 = r_2$, $u_3 = r_1 r_2$, $u_4 = r_2 r_1 r_2$, $u_5 = r_1 r_2 r_1 r_2$, and $u_6 = r_0 r_1 r_2 r_1 r_2$. We have

(a) $\mathcal{D} \cap c = \{u(i,i,0) \mid 1 \leq i \leq 6\}$

(b) $u(i,j,k) \underset{L}{\sim} u(m,n,k')$ if and only if $j = n$.

$\quad u(i,j,k) \underset{R}{\sim} u(m,n,k')$ if and only if $i = m$.

Let $Y = \{1, 2, 3, 4, 5, 6\}$ and let F_c act on Y trivially. Then the bijection

$$\pi: c \xrightarrow{\sim} \text{the set of isomorphism classes of irreducible } F_c\text{-v.b. on } Y \times Y$$

defined by $u(i, j, k) \to V(k)_{(i,j)}$ satisfies 11.2(i) and 11.2(ii). The proof is similar to that in (C).

From now on (in the proof) we assume that G is of type B_2. Then W is of type $\tilde{B}_2$.

(E). Case $c = c_2$. We have $F_c = \mathbb{Z}/(2) \times SL_2(\mathbb{C})$ and

$$c = \{\omega^p v(i, j, k) \mid p = 0, 1; \; i, j = 1, 2, 3, 4; \; k \in \mathbb{N}\}$$

where $v(i, j, k) = v_i r_0 r_2 (r_1 r_2 r_0)^k v_j^{-1}$, $v_1 = e$, $v_2 = r_1$, $v_3 = r_2 r_1$, and $v_4 = r_0 r_1$. We have

(a) $\mathcal{D} \cap c = \{v(i, i, 0) \mid 1 \le i \le 4\}$.

(b) $\omega^p v(i, j, k) \underset{L}{\sim} \omega^{p'} v(m, n, k')$ if and only if $j = n$,

$\omega^p v(i, j, k) \underset{R}{\sim} \omega^{p'} v(m, n, k')$ if and only if $i = m$.

Let $Y = \{1, 2, 3, 4\}$ and let F_c act on Y trivially. As the same way in (C), we know that the bijection

$$\pi: c \xrightarrow{\sim} \text{the set of isomorphism classes of irreducible } F_c\text{-v.b. on } Y \times Y$$

defined by $\omega^p v(i, j, k) \to (\varepsilon^p, V(k))_{(i,j)}$ satisfies 11.2(i) and 11.2(ii), where ε is the sign representation of $\mathbb{Z}/(2)$.

(F). Case $c = c_1$. We have $F_c = \mathbb{Z}/(2) \ltimes \mathbb{C}^*$, where $\mathbb{Z}/(2)$ acts on $\mathbb{C}^*$ by $z \to z^{-1}$;

$$c = \{\omega^p (r_0 r_1 \omega)^k \omega^{p'}, \omega^p (r_0 r_1 \omega)^{k'} r_0 \omega^{p'}, \omega^p (r_1 r_0 \omega)^k \omega^{p'},$$
$$\omega^p (r_1 r_0 \omega)^{k'} r_1 \omega^{p'}, \omega^p r_0 r_1 r_0 \omega^{p'} \mid p, p' = 0, 1; \; k > 0, \; k' \in \mathbb{N}\}$$

We shall regard $\mathbf{q}^i$ ($i \in \mathbb{Z}$) as the simple representation of $\mathbb{C}^*$ defined by $z \to z^i$. Let $\sigma(k)$ ($k > 0$) be the simple representation of F_c such that the restriction to $\mathbb{C}^*$ of σ_k is the direct sum of $\mathbf{q}^k$ and $\mathbf{q}^{-k}$. The sign representation ε of $\mathbb{Z}/(2)$ gives rise to a simple representation of F_c via the natural homomorphism $F_c \to \mathbb{Z}/(2)$, we denote it again by ε. Let $\sigma(0) = 1$ be the unit representation of F_c.

Let $Y = \{1, 2, 3, 4\}$ be the F_c-set such that as F_c-sets we have $\{1\} \simeq \{2\} \simeq F_c/F_c$ and $\{3, 4\} \simeq F_c/F_c^0$. As in (C), we can check that the following bijection π has the properties 11.2(i) and 11.2(ii).

$$\pi: c \xrightarrow{\sim} \text{the set of irreducible } F_c\text{-v.b. on } Y \times Y \text{ (up to isomorphisms)},$$

$$r_0 r_1 r_0 \to \mathcal{E}_{(1,1)}, \quad \omega r_0 r_1 r_0 \to \mathcal{E}_{(2,1)}, \quad r_0 r_1 r_0 \omega \to \mathcal{E}_{(1,2)}, \quad r_2 r_1 r_2 \to \mathcal{E}_{(2,2)},$$

$$(r_0 r_1 \omega)^k r_0 \to \sigma(k)_{(1,1)}, \qquad (r_0 r_1 \omega)^k r_0 \omega \to \sigma(k)_{(1,2)},$$
$$\omega(r_0 r_1 \omega)^k r_0 \to \sigma(k)_{(2,1)}, \qquad \omega(r_0 r_1 \omega)^k r_0 \omega \to \sigma(k)_{(2,2)},$$
$$(r_1 r_0 \omega)^k r_1 \to \mathbf{q}^k_{(3,3)}, \qquad \omega(r_1 r_0 \omega)^k r_1 \omega \to \mathbf{q}^{-k}_{(3,3)},$$
$$(r_1 r_0 \omega)^k r_1 \omega \to \mathbf{q}^{k-1}_{(3,4)}, \qquad \omega(r_1 r_0 \omega)^k r_1 \to \mathbf{q}^{-1-k}_{(3,4)},$$
$$(r_0 r_1 \omega)^{k'} \to \mathbf{q}^{k'}_{(1,3)}, \qquad (r_0 r_1 \omega)^{k'} \omega \to \mathbf{q}^{1-k'}_{(1,3)},$$
$$\omega(r_1 r_0 \omega)^{k'} \omega \to \mathbf{q}^{-k'}_{(3,1)}, \qquad (r_1 r_0 \omega)^{k'} \omega \to \mathbf{q}^{k'-1}_{(3,1)},$$
$$\omega(r_0 r_1 \omega)^{k'} \omega \to \mathbf{q}^{1-k'}_{(2,3)}, \qquad \omega(r_0 r_1 \omega)^{k'} \to \mathbf{q}^{k'}_{(2,3)},$$
$$(r_1 r_0 \omega)^{k'} \to \mathbf{q}^{k'-1}_{3,2}, \qquad \omega(r_1 r_0 \omega)^{k'} \to \mathbf{q}^{-k'}_{(3,2)},$$

where $k \geq 0$, $k' > 0$ are integers.

Application to Representation

11.3 We shall apply the idea in 5.13 to classify the simple $\mathbf{H}_q$-modules under the assumption that W is of type $\tilde{G}_2$ or $\tilde{B}_2$. When W is of type $\tilde{A}_1, \tilde{A}_2$, see [X3]. The results in [X3] and in this chapter show that the map $(\phi_q)_{*,c}$ (see 5.13 for its definition) is a good way to understand the classification of simple $\mathbf{H}_q$-modules even if q is a root of 1. Our second main result in the chapter is the following. We refer to chapter 5, especially 5.13, for notations.

Theorem 11.4. *Let W be an extended affine Weyl group type $\tilde{G}_2$ or $\tilde{B}_2$ as in 11.1.*

(i). Assume that E is a simple $\mathbf{J}$-module, then E_q has at most one simple constituent L such that $c_L = c_E$.

(ii). Given two simple $\mathbf{J}$-modules E and E' such that E_q has a simple constituent L with $c_L = c_E$ and E'_q has a simple constituent L' with $c_{L'} = c_{E'}$. Then $L \simeq L'$ if and only if $E \simeq E'$.

(iii). The set $\Lambda = \{(\phi_q)_{*,c}(E) \mid c$ a two-sided cell of W, E a simple $\mathbf{J}_c$-module (up to isomorphisms)$\} - \{0\}$ is a basis of $K(\mathbf{H}_q)$.

(iv). $(\phi_q)_*$ is an isomorphism if and only if $\sum_{w \in W_0} q^{l(w)} \neq 0$.

Proof. Since $(\phi_q)_*$ is surjective (see 5.13), we know that (iii) follows from (i) and (ii).

Let c be a two-sided cell of W. Each $\mathbf{J}_c$-module E gives rise to an $\mathbf{H}_q$-module E_q via the homomorphism $\phi_{q,c} \colon \mathbf{H}_q \to \mathbf{J} \to \mathbf{J}_c$. Then the assertions (i) and (ii) are equivalent to the following assertion:

(▼) Given a simple $\mathbf{J}_c$-module E, the $\mathbf{H}_q$-module E_q has at most one simple constituent L such that $c_L = c$. Suppose that the simple $\mathbf{J}_c$-module E (resp. E'), E_q (resp. E'_q) has a simple constituent L (resp. L') such that $c_L = c$ (resp. $c_{L'} = c$), then $L \simeq L'$ if and only if $E \simeq E'$.

We prove ($\blacktriangledown$) case by case. The following fact will be used repeatedly. Let M_1 and M_2 be two finite dimensional $\mathbf{H}_q$-modules. Assume that M_1 and M_2 are isomorphic, then for any h in $\mathbf{H}_q$ the eigenpolynomials of h on M_1 and on M_2 are equal (up to a scalar).

(A). When $c = c_e$, the map $(\phi_q)_{*,c}$ is an isomorphism, what we need is trivial. When $c = c_0$, in [X3] we have shown that Λ_{c_0} is a complete set of irreducible H_q-modules attached to c_0 and that $(\phi_q)_{*,c}$ is an isomorphism if and only if $\sum_{w \in W_0} q^{l(w)} \neq 0$. In (B-D) we assume that W is of type $\tilde{G}_2$.

(B). Assume that $c = c_1$. The based ring $\mathbf{J}_c$ has four simple modules E_1, E_2, E_3, E_4. We have $\dim E_1 = \dim E_2 = 3$, $\dim E_3 = 2$ and $\dim E_4 = 1$. When $q + 1 \neq 0$, one verifies that $E_{i,q}$ ($1 \leq i \leq 4$) has a unique simple constituent L_i such that $c_{L_i} = c_1$ and L_i is not isomorphic to L_j if $i \neq j$. The $\mathbf{H}_q$-module L_i ($1 \leq i \leq 4$) in fact is a quotient module of $E_{i,q}$. When $q + 1 = 0$, one can check that $E_{i,q}$ ($1 \leq i \leq 3$) has a unique simple constituent L_i such that $c_{L_i} = c_1$ and L_i is not isomorphic to L_j if $i \neq j$. We also have $(\phi_q)_{*,c}(E_4) = 0$.

(C). Assume that $c = c_2$. We have $\mathbf{J}_c \simeq M_{6 \times 6}(\mathbf{R}_{F_c})$ (the 6×6 matrix ring over $\mathbf{R}_{F_c}$, where $\mathbf{R}_{F_c} = \mathbb{C}$ tensors with the representations ring of F_c, see 11.2 and 5.4). For each semisimple conjugacy class s of $F_c = SL_2(\mathbb{C})$, we have a simple representation ψ_s of $\mathbf{J}_c$:

$$\psi_s \colon \mathbf{J}_c \simeq M_{6 \times 6}(\mathbf{R}_{F_c}) \to M_{6 \times 6}(\mathbb{C}),$$
$$(m_{ij}) \to (tr(s, m_{ij})).$$

Each simple representation of $\mathbf{J}_c$ is isomorphic to some ψ_s. Let E_s be a simple $\mathbf{J}_c$-module which provides the representation ψ_s.

The $\mathbf{H}_q$-module $E_{s,q}$ in fact is an $\mathbf{H}_q^{\geq c}$-module, where $\mathbf{H}_q^{\geq c}$ is the quotient algebra of $\mathbf{H}_q$ modulo the two-sided ideal generated by C_w, $w \underset{LR}{\leq} r_0 r_2$ but $w \notin c$. We denote the image in $\mathbf{H}_q^{\geq c}$ of C_w again by C_w. By (h) and (j) in the part (C) of the proof of 11.2, we see that the two-sided ideal of $\mathbf{H}_q^{\geq c}$ spanned by C_w, $w \underset{LR}{\sim} r_0 r_2$, is generated by C_{02}. Hence for any simple constituent L of $E_{s,q}$, the attached two-sided cell of L is c if and only if $C_{02} L \neq 0$. We have

(a) $\qquad \phi_{q_0,c}(C_{02}) = \begin{pmatrix} [2]^2 & [2] & 0 & V(1) & [2]V(1) & [2]V(1) \\ & & & 0 & & \end{pmatrix}$

(b) $\qquad \phi_{q_0,c}(C_{02121}) = \begin{pmatrix} V(1) & [2]V(1) & V(1)+[2] & [2]^2 & [2] & [2] \\ & & & 0 & & \end{pmatrix}$

where $[2] = q^{\frac{1}{2}} + q^{-\frac{1}{2}}$.

By (a) we know that $E_{s,q}$ has at most one simple constituent attached to the two-sided cell c. In fact, when $D = C_{02} E_{s,q} = 0$, the $\mathbf{H}_q$-module $E_{s,q}$ obviously has no

109

simple constituents attached to the two-sided cell c. When $D \neq 0$, from (a) we see that $\dim D = 1$. Let N be the $\mathbf{H}_q$-submodule of $E_{s,q}$ generated by elements in D, then N has a maximal submodule N_0 which does not contain D, so $C_{02}N_0 = 0$. Note that $C_{02}(E_{s,q}/N) = 0$, we know that $E_{s,q}$ has at most one simple constituent L such that $c_L = c$.

If $D \neq 0$, then either $[2] \neq 0$ or $\varphi(s) = tr(s, V(1)) \neq 0$, using (a) and (b) we see that either $C_{02}D = D$ or $C_{02121}D = D$. Thus we have either $C_{02}(N/N_0) \neq 0$ or $C_{02121}(N/N_0) \neq 0$. That is, $E_{s,q}$ has a simple constituent $L_s = N/N_0$ with $c_{L_s} = c$.

From (b) we see that the eigenpolynomial of C_{02121} on L_s is $(\lambda - \varphi(s))\lambda^{b(s)}$, where $b(s) = \dim L_s - 1$. Since $s \to \varphi(s)$ defines a bijection between the set of semisimple conjugacy classes and $\mathbb{C}$, we know that when $C_{02}E_{s,q} \neq 0$ and $C_{02}E_{t,q} \neq 0$, then $L_s \simeq L_t$ if and only if $s = t$, i.e. $E_s = E_t$.

The ($\blacktriangledown$) is proved for the two-sided cell c_2 in W.

We also showed that $(\phi_q)_{*,c}$ is an isomorphism when $q + 1 \neq 0$ and $(\phi_q)_{*,c}(E_{s,q}) = 0$ if and only if $q + 1 = 0$, $\varphi(s) = 0$. It is known that only one semisimple conjugacy class s in $SL_2(\mathbb{C})$ such that $\varphi(s) = 0$.

(D). Assume that $c = c_3$. We have $\mathbf{J}_c \simeq M_{6 \times 6}(\mathbf{R}_{F_c})$ (cf. 11.2) and

(c) $\phi_{q,c}(C_{010}) =$

$$\begin{pmatrix} [2]^3 - [2] & [2]V(1) + [2]^2 & [2]^2 V(1) + [2] & V(1)^2 + [2]V(1) & [2]V(2) & [2]^2 V(2) \\ & & 0 & & & \end{pmatrix}$$

(d) $C_{0102}C_{010} = [2]V(1) + [2]^2$

(e) $C_{010212}C_{101} = C_{010210210} + [2]C_{010210} + C_{010}$

(f) $C_{010210212}C_{010} = C_{010210210210} + [2]C_{010210210} + 2C_{010210} + [2]C_{010}$

Note that $F_c = SL_2(\mathbb{C})$. For a semisimple conjugacy class s of F_c, let ψ_s and E_s be as in (C). Via $\phi_{q,c}$, the $\mathbf{J}_c$-module E_s gives rise to an $\mathbf{H}_q$-module $E'_{s,q}$. As the same way in (C), we know that $E'_{s,q}$ has at most one simple constituent attached to the two-sided cell c. When $C_{010}E'_{s,q} = 0$, the $\mathbf{H}_q$-module $E'_{s,q}$ has no simple constituents attached to the two-sided cell c. Moreover, if $C_{010}E'_{s,q} \neq 0$, then $E'_{s,q}$ has a unique simple constituent L'_s such that $c_{L'_s} = c$. Let $b'(s) = \dim L'_s - 1$, from (c-f) we know that the eigenpolynomials of C_{0102}, C_{010212} and $C_{010210212}$ on L'_s are $(\lambda - [2]\varphi(s) - [2]^2)\lambda^{b'(s)}$, $(\lambda - \varphi(s)^2 - [2]\varphi(s))\lambda^{b'(s)}$ and $(\lambda - \varphi(s)^3 - [2]\varphi(s)^2)\lambda^{b'(s)}$, respectively. Thus if both $C_{010}E'_{s,q}$ and $C_{010}E'_{t,q}$ are not equal to zero, then $L'_s \simeq L'_t$ if and only if $s = t$, i.e. $E_s = E_t$. We have proved ($\blacktriangledown$) for $c = c_3$.

It is obvious that $C_{010}E'_{s,q} = 0$ if and only if $q + 1 = 0$ and $\varphi(s) = 0$, or $q^2 + q + 1 = 0$ and $\varphi(s) + [2] = 0$. Thus $(\phi_q)_{*,c}$ is an isomorphism if $[2]([2]^2 - 1) \neq 0$. When $[2] = 0$ or $[2]^2 = 1$, there exists a unique semisimple conjugacy class s in G such that $(\phi_q)_{*,c}(E_s) = 0$.

In (E-F) we assume that W is of type $\tilde{B}_2$.

(E). Assume that $c = c_2$. We have $\mathbf{J}_c \simeq M_{4\times 4}(\mathbf{R}_{F_c})$ and $F_c = \mathbb{Z}/(2) \times SL_2(\mathbb{C})$ (see 11.2). We have

(g) $\qquad \phi_{q,c}(C_{02}) = \begin{pmatrix} [2]^2\ V(1) + [2]\ [2]V(1)\ [2]V(1) & 0 \\ 0 & \end{pmatrix}$

(h) $\qquad \phi_{q,c}(C_{021}) = \begin{pmatrix} V(1) + [2]\ [2]V(1) + [2]^2\ V(1) + [2]\ V(1) + [2] & 0 \\ 0 & \end{pmatrix}$

(i) $\qquad \phi_{q,c}(C_\omega) = \begin{pmatrix} \varepsilon & & & 0 \\ & \varepsilon & & \\ & & \varepsilon & \\ 0 & & & \varepsilon \end{pmatrix}$

where ε is the sign representation of $\mathbb{Z}/(2)$ and we also write ε instead of $(\varepsilon, V(0))$.

For each semisimple conjugacy class s of F_c, let ψ_s be the simple representation of $\mathbf{J}_c$ defined by

$$\psi_s(m_{ij}) = (tr(s, m_{ij})) \in M_{4\times 4}(\mathbb{C}).$$

Let E_s be a simple $\mathbf{J}_c$-module which provides ψ_s. Each simple $\mathbf{J}_c$-module is isomoprhic to some E_s. Via $\phi_{q,c}$, the $\mathbf{J}_c$-module E_s gives rise to an $\mathbf{H}_q$-module $E_{s,q}$.

As the same way in (C) we know that $E_{s,q}$ has no simple constituents attached to the two-sided cell c when $C_{02}E_{s,q} = 0$, and $E_{s,q}$ has exactly one simple constituent L_s such that $c_{L_s} = c$ when $C_{02}E_{s,q} \neq 0$. Moreover, the eigenpolynomials of C_ω and C_{021} on L_s are $(\lambda - \varphi'(s))^{b(s)+1}$ and $(\lambda - \varphi(s) - [2])\lambda^{b(s)}$, respectively, where $\varphi'(s) = tr(s, \varepsilon)$, $\varphi(s) = tr(s, V(1))$ and $b(s) = \dim L_s - 1$. Therefore if both $C_{02}E_{s,q}$ and $C_{02}E_{t,q}$ are not equal to zero, then $L_s \simeq L_t$ if and only if $s = t$. We have proved the ($\blacktriangledown$) for the two-sided cell $c = c_2$.

It is obvious that $C_{02}E_{s,q} = 0$ if and only if $q + 1 = 0$, $\varphi(s) = 0$. Hence $(\phi_q)_{*,c}$ is an isomorphism when $q + 1 \neq 0$, and there exist two semisimple conjugacy classes s_1, s_2 such that $(\phi_q)_{*,c}(E_{s_i}) = 0$ $(i = 1, 2)$ when $q + 1 = 0$.

(F). Assume that $c = c_1$. We have $F_c = \mathbb{Z}/(2) \ltimes \mathbb{C}^*$, where $\mathbb{Z}/(2)$ denotes the quotient ring of $\mathbb{Z}$ moduloing the prime ideal generated by 2 and the non trivial element in $\mathbb{Z}/(2)$ acts on $\mathbb{C}^*$ by $z \to z^{-1}$. Any element in F_c is semisimple. Let $e \neq \alpha \in \mathbb{Z}/(2)$, then $\alpha z = z^{-1}\alpha$, and the set $s(\alpha) = \{\alpha z \mid z \in \mathbb{C}^*\}$ is the conjugacy class containing α. For any z in $\mathbb{C}^*$, let $s(z)$ be the conjugacy class containing z. Then $s(1) = \{1\}$, $s(-1) = \{-1\}$, $s(z) = \{z, z^{-1}\}$ if $z^2 \neq 1$. By 11.2 and 5.4 we know that

$$\{E_{s(\alpha)}, E_{s(1),\varepsilon}, E_{s(-1),\varepsilon}, E_{s(z)} \mid z \in \mathbb{C}^*\}$$

is a complete set of simple $\mathbf{J}_c$-modules, where ε is the sign representation of $\mathbb{Z}/(2)$. It is not difficult to see that $E_{s(\alpha),q}$ is a simple $\mathbf{H}_q$-module and $C_0E_{s(\alpha),q} \neq 0$ when $q + 1 \neq 0$, $(\phi_q)_{*,c}(E_{s(\alpha)}) = 0$ when $q + 1 = 0$. We always have $C_1E_{s(\alpha),q} = 0$.

111

One verifies that each $E_{s(z),q}$ ($z \in \mathbb{C}^*$) has exactly one simple constituent attached to the two-sided cell c, denote by $L_{s(z)}$. The $\mathbf{H}_q$-module $L_{s(z)}$ in fact is a quotient module of $E_{s(z),q}$. The eigenpolynomial of $C_{r_1 r_{0w}}$ on $L_{s(z)}$ is $(\lambda - z - z^{-1})\lambda^{b(z)}$, where $b(z) = \dim L_{s(z)} - 1$. So $L_{s(z)} \simeq L_{s(z')}$ if and only if $s(z) = s(z')$ when $z, z' \in \mathbb{C}^*$. One checks that for any $z \in \mathbb{C}^*$, $C_1 L_{s(z)}$ is not equal to zero if $q + 1 \neq 0$. Thus for any $z \in \mathbb{C}^*$, the $\mathbf{H}_q$-modules $L_{s(z)}$ and $E_{s(\alpha),q}$ are not isomorphic when $q + 1 \neq 0$.

We have $\dim E_{s(i),\varepsilon,q} = 1$ for $i = \pm 1$, and C_0, C_2, C_1, C_ω acts on $E_{s(i),\varepsilon,q}$ by scalars 0, 0, $[2]$, i, respectively. So $(\phi_q)_{*,c}(E_{s(i),\varepsilon}) = 0$ if $q + 1 = 0$ and $(\phi_q)_{*,c}(E_{s(i),\varepsilon}) = E_{s(i),\varepsilon,q} = L_{i,\varepsilon}$ if $q + 1 \neq 0$. Now assume that $q + 1 \neq 0$. Obviously we have $L_{1,\varepsilon} \not\simeq L_{-1,\varepsilon}$ and $L_{i,\varepsilon} \not\simeq E_{s(\alpha),q}$ ($i = \pm 1$). It is easy to see that $C_1 C_0 L_{s(z)} \neq 0$ ($z \in \mathbb{C}^*$), so $L_{s(z)}$ is not isomorphic to $L_{i,\varepsilon}$ ($z \in \mathbb{C}^*$, $i = \pm 1$). The ($\blacktriangledown$) is proved for $c = c_1$.

In the above discussion we see that $(\phi_q)_{*,c}$ is an isomorphism when $q + 1 \neq 0$ and rank $\ker(\phi_q)_{*,c} = 3$ when $q + 1 = 0$.

(G). Since $(\phi_q)_*$ is an isomorphism if and only if $(\phi_q)_{*,c}$ is an isomorphism for any two-sided cell c of W, we see that 11.4(iv) is true according to (A-F).

Theorem 11.4 is proved.

11.5. In the proof of 11.4 we have determined the kernel of $(\phi_q)_{*,c}$ explicitly for each two-sided cell c in $W - c_0$. Now we determine the kernel of $(\phi_q)_{*,c_0}$ in a different way from that in chapter 7. The kernel is 0 when $\sum_{w \in W_0} q^{l(w)} \neq 0$.

We denote w_0 the longest element of W_0. Let x_i ($i = 1, 2$) be the i-th basic dominant weight in X, then $r_i x_j = x_j r_i \in c_0$ ($i \neq j \in \{1, 2\}$), $x_1 x_2 \in c_0$. Moreover for type B_2 we have $x_1 = r_0 r_1 r_2 r_1$ and $x_2 = w r_2 r_1 r_2$ (the simple root α_1 corresponding to r_1 is long and the simple root α_2 corresponding to r_2 is short). For type G_2 we have $x_1 = r_0 r_1 r_2 r_1 r_2 r_1$ and $x_2 = r_0 r_1 r_2 r_1 r_0 r_2 r_1 r_2 r_1 r_2$ (the simple root α_1 corresponding to r_1 is short and the simple root α_2 corresponding to r_2 is long).

In $\mathbf{H}_q$ we have

(a) Case $\tilde{B}_2$: $C_{w_0} C_{x_1 r_2} = [2](C_{w_0 x_1} + ([2]^2 - 1)C_{w_0})$,
$C_{w_0} C_{x_2 r_1} = [2] C_{w_0 x_2}$, $\quad C_{w_0} C_{x_1 x_2} = C_{w_0 x_1 x_2} + [2]^2 C_{w_0 x_2}$

(b) Case $\tilde{G}_2$: $C_{w_0} C_{x_1 r_2} = [2](C_{w_0 x_1} + [5]C_{w_0})$,
$C_{w_0} C_{x_2 r_1} = [2](C_{w_0 x_2} + ([2]^2 - 1)C_{w_0 x_1} + C_{w_0})$,
$C_{w_0} C_{x_1 x_2} = C_{w_0} U_{x_1} U_{x_2} + ([2]^2 - 1)C_{w_0} U_{x_1}^2 + [5]C_{w_0} U_{x_2} + [2]^2 C_{w_0} U_{x_1} + [5]C_{w_0}$

where $[5] = q^2 + q + 1 + q^{-1} + q^{-2}$, and U_{x_1}, U_{x_2} are defined as in 2.2.

Let $V(x_1)$, $V(x_2)$ and $V(x_1 x_2)$ be the simple G-modules of highest weight x_1, x_2 and $x_1 x_2$, respectively. Then $s \to \varphi(s) = (\lambda_1, \lambda_2)$ defines a bijection between the set of semisimple conjugacy classes of G and $\mathbb{C}^2$, where $\lambda_i = tr(s, V(x_i))$ ($i = 1, 2$). Let E_s be the simple $\mathbf{J}_{c_0}$-module corresponding to a semisimple conjugacy class s of G (see [X3]). According to [X3, 3.9], (a) and (b) we have the following results.

112

Case $\tilde{B}_2$:

(c1). When $q^2 + 1 = 0$, $(\phi_q)_{*,c_0}(E_s) = 0$ if and only if $\varphi(s) = (-1, 0)$.

(c2). When $q + 1 = 0$, $(\phi_q)_{*,c_0}(E_s) = 0$ if and only if $\lambda_1 \lambda_2 = \lambda_2$, i.e. $tr(s, V(x_1 x_2)) = 0$.

(c3). When $(q+1)(q^2+1) \neq 0$, we know that $(\phi_q)_*$ is an isomorphism.

Case $\tilde{G}_2$:

(d1). When $q^4 + q^2 + 1 = 0$, $(\phi_q)_{*,c_0}(E_s) = 0$ if and only if $\varphi(s) = (q^3, q^3)$.

(d2). When $q + 1 = 0$, $(\phi_q)_{*,c_0}(E_s) = 0$ if and only if $\lambda_1 \lambda_2 - \lambda_1^2 + \lambda_2 + 1 = 0$, i.e. $tr(s, V(x_1 x_2)) = 0$.

(d3). When $(q+1)(q^4 + q^2 + 1) \neq 0$, we know that $(\phi_q)_*$ is an isomorphism.

11.6. Now we assume that W is an arbitrary irreducible affine Weyl group. Let e_n be the largest exponent of W_0. When q is a primitive $(e_n + 1)$-th root of 1, then rank $\ker(\phi_q)_{*,c_0} = 1$ (see [X3]). It is likely that rank $\ker(\phi_q)_* = 1$ in this case, i.e. $(\phi_q)_{*,c}$ is an isomorphism when $c \neq c_0$.

11.7. Relations between various $\mathbf{H}_q$. Let G be a simple algebraic group over $\mathbb{C}$. Let W_0 be its Weyl group and W be its extended affine Weyl group. Let $\mathbf{H}_{q,W_0}$ be the Hecke algebra over $\mathbb{C}$ of W_0 with parameter $q \in \mathbb{C}^*$. When $\sum_{w \in W_0} q^{l(w)} \neq 0$, It is known that there are natural isomorphisms of $\mathbb{C}$-algebras

$$\mathbf{H}_{q,W_0} \simeq \mathbb{C}[W], \qquad \mathbf{H}_{q,W_0} \simeq \mathbf{J}_{W_0}, \qquad (\text{see } [L3, GU]),$$

where $\mathbf{J}_{W_0} \subset \mathbf{J}$ stands in an obvious sense.

Reall that $\mathbf{H}_q$ is the Hecke algebra over $\mathbb{C}$ of W with parameter $q \in \mathbb{C}^*$ and $\mathbf{J}$ is the asymptotic Hecke algebra of W (defined in 2.7.) The homomorphism $\phi_q : \mathbf{H}_q \to \mathbf{J}$ is injective but never surjective. Actually it is impossible to find an isomorphism between $\mathbf{H}_q$ and $\mathbf{J}$ for any $q \in \mathbb{C}^*$. Now we would like to look the relations between $\mathbf{H}_q$ and $\mathbf{H}_1 = \mathbb{C}[W]$.

When G is of type A_1, W is of type $\tilde{A}_1$. Let s, t be the simple reflections in W. When $q + 1 \neq 0$, there is a unique isomorphism of $\mathbb{C}$-algebra between $\mathbf{H}_q$ and $\mathbb{C}[W]$ such that

$$T_s \to \frac{q+1}{2}s + \frac{q-1}{2}, \qquad T_t \to \frac{q+1}{2}t + \frac{q-1}{2}.$$

$\mathbf{H}_{-1}$ has two (resp. one) simple modules of dimension 1 when G is simply connected (resp. adjoint), $\mathbf{H}_1$ has four simple modules of dimension 1, so $\mathbf{H}_{-1} \not\simeq \mathbb{C}[W] = \mathbf{H}_1$.

Now we assume that $G = SL_3(\mathbb{C})$, the simply connected, simple algebraic group over $\mathbb{C}$ of type A_2. We shall show that

(a) $\mathbf{H}_q \not\simeq \mathbb{C}[W]$ whenever $q \neq 1$.

The extended affine Weyl group $W = \Omega \ltimes W'$ has three two-sided cells: $c_e = \Omega$, $c_1 = \{w \in W \mid a(w) = 1\}$, $c_0 = \{w \in W \mid a(w) = 3\}$. We have (see [X2, X3]) (b). $\mathbf{J}_{c_0} \simeq M_{6\times 6}(\mathbf{R}_G)$, $\mathbf{J}_{c_1} \simeq M_{3\times 3}(\mathbf{A})$, $\mathbf{J}_{c_e} \simeq \mathbb{C}[\Omega]$. Note that $\mathbf{A} = \mathbb{C}[\mathbf{q}, \mathbf{q}^{-1}]$.

Each $\mathbf{J}_c$-module E gives rise to an $\mathbf{H}_q$-module E_q via the homomorphism

$$\phi_{q,c} \colon \mathbf{H}_q \to \mathbf{J} \to \mathbf{J}_c,$$

where c is a two-sided cell of W. Each $\mathbf{J}$-module E gives rise to an $\mathbf{H}_q$-module via the homomorphism $\phi_q \colon \mathbf{H}_q \to \mathbf{J}$. Note that $\mathbf{J} = \mathbf{J}_{c_e} \oplus \mathbf{J}_{c_1} \oplus \mathbf{J}_{c_0}$.

We recall (see 5.9(e) and [X3])

(c) Each simple $\mathbf{H}_q$-module L is a quotient module of E_q for some simple $\mathbf{J}$-module E with $c_E = c_L$.

(d) Assume that E is a simple $\mathbf{J}$-module, then E_q has at most one simple constituent L such that $c_L = c_E$.

(e) Given two simple $\mathbf{J}$-modules E and E' such that E_q has a simple constituent L with $c_L = c_E$ and E_q' has a simple constituent L' with $c_{L'} = c_{E'}$. Then $L \simeq L'$ if and only if $E \simeq E'$.

For each semisimple element s of G, we have a simple $\mathbf{J}_{c_0}$-module E_s obtained through the simple representation ψ_s of $\mathbf{J}_{c_0}$:

$$\psi_s \colon \mathbf{J}_{c_0} \simeq M_{6\times 6}(\mathbf{R}_G) \to M_{6\times 6}(\mathbb{C}),$$
$$(m_{ij}) \to (tr(s, m_{ij})).$$

It is known that $E_s \simeq E_t$ if and only if s and t are conjugate in G.

For each element $a \in \mathbb{C}^*$, we have a simple $\mathbf{J}_{c_1}$-module E_a obtained through the simple representation ψ_s of $\mathbf{J}_{c_1}$:

$$\psi_a \colon \mathbf{J}_{c_1} \simeq M_{3\times 3}(\mathbf{A}) \to M_{3\times 3}(\mathbb{C}),$$
$$\text{specialize } \mathbf{q} \text{ to } a.$$

It is obvious that $E_a \simeq E_b$ if and only if $a = b$.

(f) $E_{s,q}$ ($s \in G$ semisimple) has a simple constituent $L_{s,q}$ such that $c_{L_{s,q}} = c_0$ if and only if $\mathbf{g}_{s,q} = \mathcal{N}_{s,q}$ or $q = 1$. Moreover

$$\dim L_{s,q} = \begin{cases} 6, & \text{if } \mathcal{N}_{s,q} = \{0\}, \\ 3, & \text{if } \mathcal{N}_{s,q} \neq \{0\} \text{ and } \mathcal{N}_{s,q} \text{ doesnot contain} \\ & \quad \text{any regular nilpotent element of } \mathbf{g}, \\ 1, & \text{if } \mathcal{N}_{s,q} \text{ contains some regular nilpotent elements of } \mathbf{g}. \end{cases}$$

(g) $E_{a,q}$ has a unique simple constituent $L_{a,q}$ such that $c_{L_{a,q}} = c_1$ for any $a \in \mathbb{C}^*$. Moreover (see 12.3(A))

$$\dim L_{a,q} = \begin{cases} 3, & \text{if } (a + q^{\frac{1}{2}}\xi^i)(a + q^{-\frac{1}{2}}\xi^i) \neq 0, \\ 2, & \text{if } (a + q^{\frac{1}{2}}\xi^i)(a + q^{-\frac{1}{2}}\xi^i) = 0 \text{ and } q^2 + q + 1 \neq 0, \\ 1, & \text{if } (a + q^{\frac{1}{2}}\xi^i)(a + q^{-\frac{1}{2}}\xi^i) = 0 \text{ and } q = \xi^k \neq 1, \end{cases}$$

where $q^{\frac{1}{2}}$ is a square root of q and ξ is a primitive 3-rd root of 1.

(h) For any simple $\mathbf{J}_{c_e}$-module E, the $\mathbf{H}_q$-module E_q is simple. We always have $\dim E_q = 1$.

By (f-h) we see that $\mathbb{C}[W]$ has three simple modules of dimension 2. But $\mathbf{H}_q$ has six simple modules of dimension 2 when $q^3 - 1 \neq 0$ and has no simple modules of dimension 2 when $q^2 + q + 1 = 0$. Now the assertion (a) follows. I donot know whether $\mathbf{H}_q \simeq \mathbf{H}_p$ when $(q^3 - 1)(q^2 - 1) \neq 0$ and $(p^3 - 1)(p^2 - 1) \neq 0$.

Now assume that W is arbitrary, it is likely that $\mathbf{H}_q \not\simeq \mathbf{H}_1$ whenever $q \neq 1$ and W is not of type $\tilde{A}_1 \times \cdots \tilde{A}_1$. It seems interesting to find relations among various $\mathbf{H}_q$ ($q \in \mathbb{C}^*$). Of course we have $\mathbf{H}_q \simeq \mathbf{H}_{q^{-1}}$ by 1.6(e).

12. Simple Modules Attached to c_1

Let G be a simply connected, simple complex algebraic group of rank n. Let W be its extended affine Weyl group (see 2.1), then $W = \Omega \ltimes W'$ for certain abelian group Ω and certain affine Weyl group W'. The second highest two-sided cell c_1 of W is described in [L4]. We have

$$c_1 = \{w \in W \mid a(w) = 1\}.$$

In this chapter we prove that the conjecture in [L14] is true for c_1, then classify the simple $\mathbf{H}_q$-modules attached to the two-sided cell c_1 and determine the dimensions of these simple $\mathbf{H}_q$-modules. From the results one can easily get the multiplicities of a simple $\mathbf{H}_q$-module in the standard modules $M_{s,N,q,\rho}$ when N is a subregular nilpotent element in the Lie algebra $\mathbf{g}$ of G. The multiplicities can be interpreted as the dimensions of certain cohomology groups (see [G3]). This chapter is based on the preprint [X5].

12.1. Denote $F_1 = F_{c_1}$ the reductive complex algebraic group attached to c_1 (cf. 5.14), then

$$
F_1 = \begin{cases}
\mathbb{C}^*, & \text{type } \tilde{A}_n \ (n \geq 2) \\
\mathbb{Z}/(2) \times \mathbb{Z}/(2), & \text{type } \tilde{B}_n \ (n \geq 3) \\
\mathbb{Z}/(2) \ltimes \mathbb{C}^*, & \text{type } \tilde{C}_n \ (n \geq 2) \\
\Omega, & \text{type } \tilde{D}_n \ (n \geq 4), \ \tilde{E}_n \ (n = 6,7,8) \\
\mathbb{Z}/(2), & \text{type } \tilde{F}_4 \\
\mathfrak{S}_3, & \text{type } \tilde{G}_2.
\end{cases}
$$

We refer to chapter 5, especially 5.13 for notations. The main result of the chapter is the following.

Theorem 12.2. *Let c_1 and F_1 be as above. Then*

(a). There exists a finite F_1-set Y and a bijection

$$\pi: c_1 \overset{\sim}{\to} \text{set of irreducible } F_1\text{-v.b. on } Y \times Y \text{ (up to isomorphisms)}$$

such that

  (i) *The $\mathbb{C}$-linear map $\pi : \mathbf{J}_1 \to \mathbf{K}_{F_1}(Y \times Y)$, $t_w \to \pi(w)$ is an algebra isomorphism (preserving the unit element), where $\mathbf{J}_1 = \mathbf{J}_{c_1}$.*

  (ii) *For each $w \in c_1$ we have $\pi(w^{-1}) = \widetilde{\pi(w)}$.*

(b). Let E_1 and E_2 be two simple $\mathbf{J}_1$-modules. Then

  (i) *The $\mathbf{H}_q$-module E_q (see 5.13 for the definition) has at most one simple constituent L such that $c_L = c_1$.*

(ii) *Suppose that E_q (resp. E_q') has a simple constituent L (resp. L') such that $c_L = c_1$ (resp. $c_{L'} = c_1$), then $L \simeq L'$ if and only if $E \simeq E'$. Thus the set $\Lambda_1 = \{(\phi_q)_{*,c_1}(E) \mid E$ a simple $\mathbf{J}_1$-module (up to isomorphisms)$\} - \{0\}$ is the set of simple $\mathbf{H}_q$-modules (up to isomorphisms) with attached two-sided cell c_1 (cf. 5.9(e)).*

The theorem supports the conjecture in [L14] and the idea in 5.13.

12.3. The rest of the chapter will be concerned with the proof of 12.2. We do it case by case. We also determine the dimensions of simple $\mathbf{H}_q$-modules attached to the two-sided cell c_1. The simple reflections will be numbered according to the Coxeter graphs in 1.3. We make some conventions.

If $s_{i_1} s_{i_2} \ldots s_{i_k}$ is a reduced express of an element $w \in W'$, we often write $i_1 i_2 \ldots i_k$, $C_{i_1 i_2 \ldots i_k}$, $t_{i_1 i_2 \ldots i_k}$ instead of w, C_w, t_w respectively. Let i, j, k be integers in $[0, n]$. When $i, k \le j$, it will be convenient to set

$$s_{i,j} := s_i s_{i+1} \cdots s_j, \qquad s_{j,i} := s_j s_{j-1} \cdots s_i,$$

$$s_{i,j,k} := s_i s_{i+1} \cdots s_j s_{j-1} \cdots s_k.$$

For a simple $\mathbf{J}_1$-module E, we also use $(\phi_q)_{*,c_1}(E)$ for the direct sum of simple constituents of E_q attached to c_1.

(A). Type $\tilde{A}_n$ $(n \ge 2)$. In this part we assume that W is of type $\tilde{A}_n$ $(n \ge 2)$ (we omit the case $\tilde{A}_1$, see [X3] for the case).

Let $\omega \in \Omega$ be such that $\omega s_n = s_0 \omega$ and $\omega s_i = s_{i+1} \omega$ for $i = 0, 1, ..., n-1$. Then we have
$$c_1 = \{w(i,j,k), \ u(i,j,k) \mid 0 \le i, j \le n, \ k \in \mathbb{N}\},$$
where $w(i,j,k) = \omega^i s_1 (\omega s_1)^k \omega^{-j}$ and $u(i,j,k) = \omega^i s_1 (\omega^{-1} s_1)^k \omega^{-j}$.

Let $Y = \{1, 2, \ldots, n, n+1\}$ and let $F_1 = \mathbb{C}^*$ act on Y trivially. We shall regard $\mathbf{q}^k$ $(k \in \mathbb{Z})$ as the simple representation $\mathbb{C}^* \to \mathbb{C}$, $z \to z^k$. It is easy to check that the following bijection

$$\pi: c_1 \tilde{\to} \text{set of irreducible } F_1\text{-v.b. on } Y \times Y \text{ (up to isomorphisms)}$$

$$w(i,j,k) \to \mathbf{q}^k_{(i+1,j+1)}, \qquad u(i,j,k) \to \mathbf{q}^{-k}_{(i+1,j+1)}$$

has the properties (i) and (ii) in 12.2(a).

The convolution algebra $\mathbf{K}_{F_1}(Y \times Y)$ is naturally isomorphic to $M_{n+1}(\mathbf{A})$, the $(n+1) \times (n+1)$ matrix ring over $\mathbf{A} = \mathbb{C}[\mathbf{q}, \mathbf{q}^{-1}]$. For each $a \in \mathbb{C}^*$, specializing $\mathbf{q}$ to a, we get a simple representation of $\mathbf{J}_1$:

$$\psi_a : \mathbf{J}_1 \to \mathbf{K}_{F_1}(Y \times Y) \to M_{n+1}(\mathbf{A}) \to M_{n+1}(\mathbb{C}).$$

117

Every simple representation of $\mathbf{J}_1$ is isomorphic to some ψ_a. Let E_a be a simple $\mathbf{J}_1$-module providing the representation ψ_a, then $\dim E_a = n + 1$. The $\mathbf{J}_1$-module E_a gives rise to an $\mathbf{H}_q$-module $E_{a,q}$ via ϕ_{q,c_1}. Let $v_1, v_2, ..., v_{n+1}$ be the natural base of $E_{a,q}$, we have

(a1) $C_\omega v_{n+1} = v_1$, and $C_\omega v_i = v_{i+1}$ for $i = 1, 2, ..., n$.

(a2) $C_1 v_1 = [2]v_1$, $C_1 v_2 = a v_1$, $C_1 v_{n+1} = a^{-1} v_1$, and $C_1 v_i = 0$ for $i = 3, ..., n$.

Where $[2] = q^{\frac{1}{2}} + q^{-\frac{1}{2}}$.

For any w in c_1 we can find $h_1, h_2 \in \mathbf{H}_q$ such that $C_w = h_1 C_1 h_2$. Using (a2) we see that $\dim C_1 E_{a,q} = 1$. Therefore $E_{a,q}$ has a unique simple constituent attached to the two-sided cell c_1, denote by L_a. The eigenpolynomial of $C_{1\omega}$ on L_a is $(\lambda - a)\lambda^{\theta(a)}$, where $\theta(a) = \dim L_a - 1$. Therefore $L_a \simeq L_b$ if and only if $a = b$. Thus 12.2(b) is proved for type $\tilde{A}_n$ $(n \geq 2)$.

Now we determine the dimension of L_a. The $\mathbf{H}_q$-module L_a in fact is the unique simple quotient module of $E_{a,q}$. Let N_a be the maximal submodule of $E_{a,q}$, then we have

$$
N_a = \begin{cases}
0, & \text{if } (a + q^{\frac{1}{2}}\xi^i)(a + q^{-\frac{1}{2}}\xi^i) \neq 0, \\
< \kappa_i >, & \text{if } (a + q^{\frac{1}{2}}\xi^i)(a + q^{-\frac{1}{2}}\xi^i) = 0, \text{ and } \sum_{m=0}^{n} q^m \neq 0, \\
< \kappa_i, \kappa_{i+k} >, & \text{if } a + q^{\frac{1}{2}}\xi^i = 0 \text{ and } q = \xi^k \neq 1, \\
< \kappa_i, \kappa_{i-k} >, & \text{if } a + q^{-\frac{1}{2}}\xi^i = 0 \text{ and } q = \xi^k \neq 1,
\end{cases}
$$

where $\kappa_i = v_1 + \xi^{-i} v_2 + \cdots + \xi^{-ni} v_{n+1}$ and ξ is a primitive $(n+1)$-th root of 1. By this we get the following fact.

$$
\dim L_a = \begin{cases}
n + 1, & \text{if } (a + q^{\frac{1}{2}}\xi^i)(a + q^{-\frac{1}{2}}\xi^i) \neq 0, \\
n, & \text{if } (a + q^{\frac{1}{2}}\xi^i)(a + q^{-\frac{1}{2}}\xi^i) = 0 \text{ and } \sum_{m=0}^{n} q^m \neq 0, \\
n - 1, & \text{if } (a + q^{\frac{1}{2}}\xi^i)(a + q^{-\frac{1}{2}}\xi^i) = 0 \text{ and } q = \xi^k \neq 1.
\end{cases}
$$

(B). Type $\tilde{B}_n$ $(n \geq 3)$. In this part we assume that $W = \Omega \ltimes W'$ is of type $\tilde{B}_n$ $(n \geq 3)$.

Let $\omega \in \Omega$ be such that $\omega s_0 = s_1 \omega$, $\omega s_1 = s_0 \omega$, and $\omega s_i = s_i \omega$ for $i = 2, ..., n$. Then the two-sided cell c_1 consists of the following elements (recall the conventions at the beginning of the proof):

$$s_0 \omega^p, \quad s_0 s_2 s_1 \omega^p, \quad s_1 s_2 s_0 \omega^p, \quad s_0 s_{2,n,2} s_0 \omega^p, \quad s_n s_{n-1} s_n \omega^p$$

$$s_0 s_{2,n,i} \omega^p \ (1 \leq i \leq n-1), \quad s_0 s_{2,i} \omega^p \ (2 \leq i \leq n), \quad s_{i,j} \omega^p \ (1 \leq i \leq j \leq n),$$

$$s_{i,j} \omega^p \ (1 \leq j \leq i \leq n), \quad s_{i,n,2} s_0 \omega^p \ (1 \leq i \leq n), \quad s_{i,2} s_0 \omega^p \ (2 \leq i \leq n),$$

$$s_{i,n,j} \omega^p \ (1 \leq i \leq n, \ 1 \leq j \leq n - 1),$$

where $p = 0, 1$.

For each element $w \in c_1$, there exist unique pair i, j such that $l(s_i w) < l(w)$ and $l(w s_j) < l(w)$. Assume that $w \omega^p \in W'$, then w is completely determined by $i, j, l(w), p$, we then write $w(i, j, k, p)$ instead of w, where $k = l(w)$.

Let $Y = \{0, 1, \ldots, n, n'\}$. We define an action of $F_1 = \mathbb{Z}/(2) \times \mathbb{Z}/(2)$ on Y by setting $ai = i$ $(0 \le i \le n - 1)$, $a \in F_1$, and $a_1 n = a_2 n = n'$, where $a_1 = (\bar{1}, \bar{0})$, $a_2 = (\bar{0}, \bar{1}) \in F_1$. Let V_i $(1 \le i \le 4)$ be the simple F_1-module such that a_1, a_2 acts on V_1 by scalar $-1, 1$; on V_2 by scalar $1, -1$; on V_3 by scalar $-1, -1$; on V_4 by scalar $1, 1$; respectively. Let V_1' and V_2' be simple $F_1' = \{e, a_1 a_2\}$-modules such that $a_1 a_2$ acts on V_i' by scalar $(-1)^i$ $(i = 1, 2)$, where $e = (\bar{0}, \bar{0}) \in F_1$.

We define a bijection

$$\pi: c_1 \xrightarrow{\sim} \text{set of irreducible } F_1\text{-v.b. on } Y \times Y \text{ (up to isomorphisms)}$$

as follows.

If $i \ne n \ne j$, we set

$$\pi(w(i, j, k, p)) = \begin{cases} V_{1(i,j)}, & \text{if } p = 1 \text{ with } k \text{ maximal,} \\ V_{2(i,j)}, & \text{if } p = 1 \text{ with } k \text{ minimal,} \\ V_{3(i,j)}, & \text{if } p = 0 \text{ with } k \text{ maximal,} \\ V_{4(i,j)}, & \text{if } p = 0 \text{ with } k \text{ minimal.} \end{cases}$$

If $i \ne j$ and $i = n$ or $j = n$, we set

$$\pi(w(i, j, k, p)) = \begin{cases} V_{1(i,j)}', & \text{if } p = 1, \\ V_{2(i,j)}', & \text{if } p = 0, \end{cases}$$

and we set

$$\pi(n) = V_{2(n,n)}', \quad \pi(n\omega) = V_{1(n,n)}', \quad \pi(n(n-1)n) = V_{2(n,n')}', \quad \pi(n(n-1)n\omega) = V_{1(n,n')}'.$$

One may check that π induces an isomorphism π of $\mathbb{C}$-algebra between $\mathbf{J}_1$ and $\mathbf{K}_{F_1}(Y \times Y)$ and $\pi(w^{-1}) = \widetilde{\pi(w)}$ if $w \in c_1$. 12.2(a) is proved in this case.

Now we consider simple $\mathbf{J}_1$-modules. There are four semisimple conjugacy classes in F_1: $e, a_1, a_2, a_1 a_2$. For any $a \in F_1$, we have $A(a) = F_1$. According to 5.4, we see that $\mathbf{J}_1$ has six simple modules (up to isomorphisms): $E_1 = E_{a_1}$, $E_2 = E_{a_2}$, $E_3 = E_{a_1 a_2}$, $E_4 = E_e$, $E_5 = E_{a_1 a_2, V_3}$ and $E_6 = E_{e, V_3}$. Via ϕ_{q, c_1}, the $\mathbf{J}_1$-module E_p $(1 \le p \le 6)$ gives rise to an $\mathbf{H}_q$-module $E_{p,q}$.

By definition, $E_{p,q}$ $(p = 1, 2)$ has a base $v_0, v_1, \ldots, v_{n-1}$ defined by $v_i : Y^{a_p} \to \mathbb{C}$, $j \to \delta_{ij}$, $(0 \le i, j \le n - 1)$, $p = 1, 2$. We have

(b1) $C_0 v_0 = [2] v_0$, $C_0 v_2 = v_0$, $C_0 v_i = 0$ $(i \ne 0, 2, 1 \le i \le n - 1)$.

(b2) $C_\omega v_0 = (-1)^{p-1} v_1$, $C_\omega v_1 = (-1)^{p-1} v_0$, $C_\omega v_i = (-1)^{p-1} v_i$ $(2 \le i \le n - 1)$.

(b3) $C_2 v_0 = C_2 v_1 = C_2 v_3 = v_2$, $C_2 v_2 = [2] v_2$, $C_i v_i = 0$ $(4 \le i \le n - 1, n \ge 4)$, and $C_2 v_0 = C_2 v_1 = v_2$, $C_2 v_2 = [2] v_2$, when $n = 3$.

119

(b4) $C_i v_{i-1} = C_i v_{i+1} = v_i$, $C_i v_i = [2]v_i$, $C_i v_j = 0$ $(0 \leq j \leq n-1$, $j \neq i, i-1, i+1$, $3 \leq i \leq n-2$, $n \geq 4)$.

(b5) $C_{n-1} v_{n-2} = v_{n-1}$, $C_{n-1} v_{n-1} = [2]v_{n-1}$, $C_{n-1} v_j = 0$ $(0 \leq j \leq n-2$, $n \geq 4)$.

We always have

(b6) $C_n E_{p,q} = 0$ $(p = 1, 2)$.

By (b1-b5) we see that $E_{p,q}$ $(p = 1, 2)$ has a unique simple constituent attached to the two-sided cell c_1, denote by L_p. We have $L_1 \not\simeq L_2$ since the eigenpolynomial of $C_{20\omega}$ on L_p $(p = 1, 2)$ is $(\lambda - (-1^p))\lambda^{\dim L_p - 1}$.

Similarly we know that $E_{p,q}$ $(p = 3, 4)$ has a unique simple constituent attached to the two-sided cell c_1, denote by L_p. We claim that L_3 is not isomorphic to L_4. In fact, let $e_i \in E_{p,q}$ $(0 \leq i \leq n)$ be defined by $e_i : Y \to \mathbb{C}, y \to \delta_{iy}$ $(y \in Y$, $1 \leq i \leq n-1)$ and $e_n(n) = e_n(n') = 1$, $e_n(j) = 0$ $(0 \leq j \leq n-1)$. Then we have

(b7) $C_0 e_0 = [2]e_0$, $C_0 e_2 = e_0$, $C_0 e_i = 0$ $(i \neq 0, 2, 1 \leq i \leq n)$.

(b8) $C_\omega e_0 = (-1)^p e_1$, $C_\omega e_1 = (-1)^p e_0$, $C_\omega v_i = (-1)^p v_i$ $(2 \leq i \leq n)$.

(b9) $C_2 e_0 = C_2 e_1 = C_2 e_3 = e_2$, $C_2 e_2 = [2]e_2$, $C_2 v_i = 0$ $(4 \leq i \leq n$, $n \geq 4)$, and $C_2 e_0 = C_2 e_1 = e_2$, $C_2 e_3 = 2e_2$, $C_2 e_2 = [2]e_2$ when $n = 3$.

(b10) $C_i e_i = [2]e_i$, $C_i e_{i-1} = C_i e_{i+1} = e_i$, $C_i e_j = 0$ $(3 \leq i \leq n-2, 1 \leq j \leq n, n \geq 4)$.

(b11) $C_{n-1} e_{n-2} = e_{n-1}$, $C_{n-1} e_n = 2e_{n-1}$, $C_{n-1} e_{n-1} = [2]e_{n-1}$, $C_{n-1} v_i = 0$, $(0 \leq i \leq n-3$, $n \geq 4)$.

(b12) $C_n e_{n-1} = e_n$, $C_n e_n = [2]e_n$, $C_n v_i = 0$, $(0 \leq i \leq n-2)$.

So the eigenpolynomial of $C_{20\omega}$ on L_p $(p = 3, 4)$ is $(\lambda + (-1)^p)\lambda^{\dim L_p - 1}$ and $L_3 \not\simeq L_4$.

We have $\dim E_{5,q} = \dim E_{6,q} = 1$ and C_i $(0 \leq i \leq n-1)$ acts on $E_{p,q}$ $(p = 5, 6)$ by scalar zero and C_n acts on $E_{p,q}$ $(p = 5, 6)$ by scalar $[2]$, C_ω acts on $E_{p,q}$ $(p = 5, 6)$ by scalar $(-1)^p$. So $(\phi_q)_{*,c_1}(E_p) = 0$ $(p = 5, 6)$ if and only if $q + 1 = 0$ and $(\phi_q)_{*,c_1}(E_p) = E_{p,q} = L_p$ $(p = 5, 6)$ is an $\mathbf{H}_q$-module attached to c_1 if $q + 1 \neq 0$. Obviously we have $L_5 \not\simeq L_6$.

By (b7), (b11-b12) we see that $L_p \not\simeq L_{p'}$ $(p = 1, 2; p' = 3, 4, 5, 6)$, we also have $L_p \not\simeq L_{p'}$ $(p = 3, 4; p' = 5, 6)$ since $C_0 L_p \neq 0$, $C_0 L_{p'} = 0$. 12.2(b) is proved in this case.

One may check that L_p is the unique simple quotient module of $E_{p,q}$ $(1 \leq p \leq 6)$. Let N_p be the maximal submodule of $E_{p,q}$, then

(b13) When $p = 1, 2$, we have

$$
N_p = \begin{cases}
0, & \text{if } (q+1)(q^{n-1}+1) \neq 0, \\
< v_0 - v_1 >, & \text{if } q+1 = 0 \text{ and } n \text{ odd}, \\
< v_0 - v_1, \sum_{i=1}^{n/2}(-1)^i v_{2i-1} >, & \text{if } q+1 = 0 \text{ and } n \text{ even}, \\
< \sum_{i=0}^{n-1} \alpha_i v_i >, & \text{if } q+1 \neq 0 \text{ and } q^{n-1}+1 = 0,
\end{cases}
$$

where $\alpha_0 = \alpha_1 = 1$, $\alpha_i = (-q)^{\frac{i-1}{2}} + (-q)^{\frac{1-i}{2}}$, $2 \leq i \leq n-1$. Thus

$$\dim L_p = \begin{cases} n, & \text{if } (q+1)(q^{n-1}+1) \neq 0, \\ n-1, & \text{if } q+1 = 0, n \text{ odd or if } q+1 \neq 0 \text{ but } q^{n-1}+1 = 0, \\ n-2, & \text{if } q+1 = 0 \text{ and } n \text{ even}. \end{cases}$$

(b14) When $p = 3, 4$, we have

$$N_p = \begin{cases} 0, & \text{if } (q+1)(q^{n-1}-1) \neq 0, \\ < e_0 - e_1 >, & \text{if } q+1 = 0 \text{ and } n \text{ even}, \\ < e_0 - e_1, \sum_{i=1}^{\frac{n+1}{2}} (-1)^i (2 - \delta_{n,2i-1}) e_{2i-1} > & \text{if } q+1 = 0 \text{ and } n \text{ odd}, \\ < \sum_{i=0}^{n} \alpha_i e_i >, & \text{if } q+1 \neq 0, q^{n-1}-1 = 0, \end{cases}$$

where $\alpha_0 = \alpha_1 = 1$, $\alpha_i = (-q)^{\frac{i-1}{2}} + (-q)^{\frac{1-i}{2}}$, $2 \leq i \leq n-1$, $\alpha_n = (-q)^{\frac{n-1}{2}}$. Thus

$$\dim L_p = \begin{cases} n+1, & \text{if } (q+1)(q^{n-1}+1) \neq 0, \\ n, & \text{if } q+1 = 0, n \text{ even or if } q+1 \neq 0 \text{ but } q^{n-1}-1 = 0, \\ n-1, & \text{if } q+1 = 0 \text{ and } n \text{ odd}. \end{cases}$$

Finally we have $\dim L_p = 1$ ($p = 5, 6$) when $q+1 \neq 0$.

(C). Type $\tilde{C}_n$ ($n \geq 2$): In this part we assume that W is of type $\tilde{C}_n$ ($n \geq 2$). Then $F_1 = \mathbb{Z}/(2) \ltimes \mathbb{C}^*$, where $\mathbb{Z}/(2)$ acts on $\mathbb{C}^*$ by $z \to z^{-1}$.

Let $\omega \in \Omega$ be such that $\omega s_i = s_{n-i} \omega$ ($0 \leq i \leq n$). For $k \in \mathbb{N}$, define

$$w(0, 0, k) = (s_{0,n-1}\omega)^k s_0, \quad w(n, n, k) = \omega w(0, 0, k)\omega,$$

$$w(0, n, k) = w(0, 0, k)\omega, \quad w(n, 0, k) = \omega w(0, 0, k).$$

For $0 < i, j < n$, $k \geq 1$ and $k' \in \mathbb{N}$, let

$$w(0, i, k) = w(0, 0, k-1)s_{1,n-i}\omega, \quad u(i, 0, k) = w(0, i, k)^{-1},$$

$$u(0, i, k') = w(0, 0, k')s_{1,i}, \quad w(i, 0, k') = u(0, i, k')^{-1},$$

$$w(n, i, k) = w(n, n, k-1)s_{n-1,i}, \quad u(i, n, k) = w(n, i, k)^{-1},$$

$$u(n, i, k') = w(n, n, k')s_{n-1,n-i}\omega, \quad w(i, n, k') = u(n, i, k')^{-1},$$

$$w(i, j, k) = s_{i,1}w(0, 0, k-1)s_{1,n-j}\omega, \quad w(i, j, 0) = s_{i,j},$$

$$u(i, j, k) = s_{i,n-1}w(n, n, k-1)s_{n-1,n-j}\omega, \quad u(i, j, 1) = s_{i,n-j}\omega,$$

$$w(i, j, k'+2) = s_{i,n-1}w(n, n, k')s_{n-1,j}, \quad w(i, j, k)' = s_{i,1}w(0, 0, k)s_{1,j}.$$

We set $w(i,j,0) = u(i,j,0)$, $w(i,j,0)' = u(i,j,0)'$, $(0 \leq i,j \leq n)$. Then c_1 consists of the following elements

$$w(i,j,k), u(i,j,k), w(i,j,k)', u(i,j,k)', 010, 010\omega, \omega010, \omega010\omega, \quad 0 \leq i,j \leq n, \; k \geq 0.$$

Let $Y = \{0,n,i,i' \,|\, 0 < i < n\}$ be an F_1-set such that as F_1-sets we have $\{0\} \simeq \{n\} \simeq F_1/F_1$ and $\{i,i'\} \simeq F_1/F_1^0$ $(0 < i < n)$. Then the bijection

$$\pi: c_1 \xrightarrow{\sim} \text{set of irreducible } F_1\text{-v.b. on } Y \times Y \text{ (up to isomorphisms)}$$

defined by

$$
\begin{aligned}
w(i,j,k) &\to \sigma(k)_{(i,j)}, & i,j &= 0,n, \\
010 &\to \varepsilon_{(0,0)}, & 010\omega &\to \varepsilon_{(0,n)}, \\
\omega010 &\to \varepsilon_{(n,0)}, & \omega010\omega &\to \varepsilon_{(n,n)}, \\
w(i,j,k) &\to \mathbf{q}_{(i,j)}^k, & u(i,j,k) &\to \mathbf{q}_{(i,j)}^{-k},
\end{aligned}
$$

$$(0 \leq i,j \leq n, \quad \text{and} \quad i \neq 0,n, \text{ or } j \neq 0,n,)$$

$$
w(i,j,k)' \to \mathbf{q}_{(i,j')}^k, \qquad u(i,j,k)' \to \mathbf{q}_{(i,j')}^{-k},
$$

$$(0 < i,j < n,)$$

satisfying (i) and (ii) in 12.2(a), where $\sigma(k)$ $(k > 0)$ is the simple representation of F_1 such that its restriction to $\mathbb{C}^*$ is the direct sum of $\mathbf{q}^k$ and $\mathbf{q}^{-k}$, and $\sigma(0) = 1$ is the unit representation of F_1, and ε is the one dimensional representation of F_1 which is not isomorphic to the unit representation of F_1. The part (ii) is obvious. To see that part (i) is true we need to check te following fact (c1).

(c1) $t_w t_u = \pi(w)\pi(u)$ for $w, u \in c_1$.

The proof is similar to part (C) in the proof of 11.2. First we have

(c2) $t_i t_w = t_w t_j = t_w$ provided that $l(s_i w) = l(w s_j) < l(w)$, and $t_u t_w = t_w t_u = 0$ provided that $l(us) > l(u)$ but $l(sw) < l(w)$ for some $s \in S$, see [L12].

(c3) It is easy to check that (c1) for $w = \omega^p 010\omega^{p'}$ $(p, p' = 0, 1)$.

Now suppose that $w = w(i,j,k)$ or $u(i,j,k)$ or $w(i,j,k)'$ or $u(i,j,k)'$, when $k = 1$ it is not difficult to verify (c1). Using this fact and (c2-c3) we see that (c1) is true. 12.2(a) is proved for type $\tilde{C}_n$ $(n \geq 2)$.

Now we prove 12.2(b). Let $\alpha \in \mathbb{Z}/(2)$ be such that $\alpha z = z^{-1}\alpha$ for any $z \in \mathbb{C}^*$. Then the conjugacy class containing α is $s(\alpha) = \{\alpha z \,|\, z \in \mathbb{C}^*\}$. For any $z \in \mathbb{C}^*$, let $s(z)$ be the conjugacy class containing z, then $s(z) = \{z, z^{-1}\}$, $s(1) = \{1\}$ and $s(-1) = \{-1\}$. According to 5.4 and (i) we see that

$$\{E_{s(\alpha)}, E_{s(1),\varepsilon}, E_{s(-1),\varepsilon}, E_{s(z)} \,|\, z \in \mathbb{C}^*\}$$

is a complete set of simple J_1-modules, where ε is the restriction to $\mathbb{Z}/(2)$ of the one dimensional representation ε of F_1.

It is easy to see that $E_{s(\alpha),q}$ is a simple $\mathbf{H}_q$-module of dimension 2 and $C_0 E_{s(\alpha),q} \neq 0$ when $q + 1 \neq 0$, and $(\phi_q)_{*,c_1}(E_{s(\alpha)}) = 0$ when $q + 1 = 0$. We always have $C_i E_{s(\alpha),q} = 0$ $(1 \leq i \leq n-1)$.

One verifies that $E_{s(z),q}$ $(z \in \mathbb{C}^*)$ has exactly one simple constituent attached to c_1, denote by $L_{s(z)}$. The eigenpolynomial of $C_{10\omega}$ on $L_{s(z)}$ is $(\lambda - z - z^{-1})\lambda^{b(z)}$, where $b(z) = \dim L_{s(z)} - 1$. So $L_{s(z)} \simeq L_{s(z')}$ if and only if $s(z) = s(z')$ when $z, z' \in \mathbb{C}^*$. One checks that for any $z \in \mathbb{C}^*$, $C_i L_{s(z)} \neq 0$ $(1 \leq i \leq n-1)$ if $q + 1 \neq 0$, thus $L_{s(z)} \not\simeq E_{s(\alpha),q}$ when $q + 1 \neq 0$ for any $z \in \mathbb{C}^*$.

We have $(\phi_q)_{*,c_1}(E_{s(p),\varepsilon}) = 0$ $(p = \pm 1)$ if and only if $n = 2$ and $q + 1 = 0$. If $(\phi_q)_{*,c_1}(E_{s(p),\varepsilon}) \neq 0$, it is easy to verify that $(\phi_q)_{*,c_1}(E_{s(p),\varepsilon}) = L_{p,\varepsilon}$ is a simple $\mathbf{H}_q$-module attached to c_1. C_0 acts on $L_{p,\varepsilon}$ by scalar 0. The eigenpolynomial of $C_{12...(n-2)\omega}$ $(n \geq 3)$ or C_ω $(n = 2)$ on $L_{p,\varepsilon}$ is $(\lambda - p)\lambda^{\dim L_{p,\varepsilon}-1}$. So we have $L_{1,\varepsilon} \not\simeq L_{-1,\varepsilon}$, $L_{p,\varepsilon} \not\simeq E_{s(\alpha),q}$ $(p = \pm 1)$, and $L_{p,\varepsilon} \not\simeq L_{s(z)}$ $(p = \pm 1, z \in \mathbb{C}^*)$.

The 12.2(b) is proved in this case.

For a simple $\mathbf{J}_1$-module E, one verifies that $(\phi_q)_{*,c_1}(E)$ is the unique simple quotient module of E_q when $(\phi_q)_{*,c_1}(E) \neq 0$. We shall determine the maximal submodule of E_q when $(\phi_q)_{*,c_1}(E) \neq 0$.

When $q + 1 \neq 0$, we have $\dim L_{s(\alpha)} = 2$. When $q + 1 \neq 0$, $n = 2$, we have $\dim L_{p,\varepsilon} = 1$ $(p = \pm 1)$. Now assume that $n \geq 3$, By definition, $L_{p,\varepsilon}$ $(p = \pm 1)$ has a base $v_1, v_2, ..., v_{n-1}$ defined by $v_i : Y \to \mathbb{C}, i \to 1, i' \to -1, y \to 0$ if $y \neq i, i'$, $y \in Y$. Let $N_{p,\varepsilon}$ be the maximal submodule of $E_{s(p),\varepsilon,q}$, then we have

$$N_{p,\varepsilon} = \begin{cases} < \sum_{i=1}^{\frac{n}{2}} (-1)^i v_{2i-1} >, & \text{if } q+1 = 0 \text{ and } n \text{ even,} \\ < \sum_{i=1}^{n-1} \alpha_i v_1 >, & \text{if } q+1 \neq 0 \text{ and } q^n = 1, \\ 0, & \text{otherwise,} \end{cases}$$

where $\alpha_i = (-q)^{\frac{i-1}{2}} + (-q)^{\frac{i-3}{2}} + \cdots + (-q)^{\frac{3-i}{2}} + (-q)^{\frac{1-i}{2}}$, $1 \leq i \leq n-1$.

So if $n \geq 3$, we have

$$\dim L_{p,\varepsilon} = \begin{cases} n-2, & \text{if } q+1 = 0 \text{ and } n \text{ even; or } q+1 \neq 0 \text{ and } q^n = 1, \\ n-1, & \text{otherwise.} \end{cases}$$

Now we consider $L_{s(z)}$ $(z = \pm 1)$. By definition, $E_{s(z),q}$ $(z = \pm 1)$ has a base $v_0, v_1, ..., v_n$ defined by $v_i : Y \to \mathbb{C}, i \to 1, i' \to 1, y \to 0$ if $y \neq i, i'$, $y \in Y$ (we set $0' = 0, n' = n$). Let $N_{s(z)}$ be the maximal submodule of $E_{s(z),q}$, then we have

$$N_{s(z)} = \begin{cases} < \sum_{i=0}^{\frac{n}{2}} (-1)^i v_{2i-1} >, & \text{if } q+1 = 0 \text{ and } n \text{ even,} \\ < \sum_{i=0}^{n} \alpha_i v_1 >, & \text{if } q+1 \neq 0 \text{ and } q^{n+1} + 1 = 0, \\ 0, & \text{otherwise,} \end{cases}$$

123

where $\alpha_i = (-q)^{\frac{i}{2}} + (-q)^{\frac{-i}{2}}$ $(0 \le i \le n-1)$, $\alpha_n = z(-q)^{\frac{n}{2}} + z(-q)^{\frac{-n}{2}}$.

Thus we have $(z = \pm 1)$

$$
\dim L_{s(z)} = \begin{cases} n, & \text{if } q+1 = 0 \text{ and } n \text{ even; or } q+1 \ne 0 \text{ and } q^{n+1} + 1 = 0, \\ n+1, & \text{otherwise.} \end{cases}
$$

For $z \in \mathbb{C}^*$, $z^2 \ne 1$, the $\mathbf{H}_q$-module $L_{s(z),q}$ has a base v_a $(a \in Y)$ defined by $v_a : Y \to \mathbb{C}$, $a \to 1$, $y \to 0$ if $y \ne a$, $y \in Y$. Let $N_{s(z)}$ be the maximal submodule of $E_{s(z),q}$, then we have

$$
N_{s(z)} = \begin{cases} < v', v'' >, & \text{if } q+1 = 0, \, n \text{ odd and } z^2 + 1 = 0, \\ < e' >, & \text{if } q+1 \ne 0 \text{ and } q^n z^2 - 1 = 0 \text{ but } q^n z^{-2} - 1 \ne 0, \\ < e'' >, & \text{if } q+1 \ne 0 \text{ and } q^n z^2 - 1 \ne 0 \text{ but } q^n z^{-2} - 1 = 0, \\ < e', e'' >, & \text{if } q+1 \ne 0 \text{ and } q^n z^2 - 1 = 0, \, q^n z^{-2} - 1 = 0, \\ 0, & \text{otherwise,} \end{cases}
$$

where

$$
v' = \sum_{i=0}^{\frac{n-1}{2}} (-1)^i (v_{2i-1} - v_{(2i-1)'}) - (-1)^{\frac{n+1}{2}} z v_n,
$$

$$
v'' = v_0 + \sum_{i=0}^{\frac{n-1}{2}} (-1)^i (v_{2i} - v_{(2i)'}),
$$

$$
e' = \sum_{i=0}^{n-1} (-q)^{\frac{i}{2}} v_i + \sum_{i=1}^{n-1} (-q)^{-\frac{i}{2}} v_{i'} + z(-q)^{\frac{n}{2}} v_n,
$$

$$
e'' = \sum_{i=0}^{n-1} (-q)^{-\frac{i}{2}} v_i + \sum_{i=1}^{n-1} (-q)^{\frac{i}{2}} v_{i'} + z(-q)^{-\frac{n}{2}} v_n.
$$

Thus for $z \in \mathbb{C}^*$ $(z \ne \pm 1)$ we have

$$
\dim L_{s(z)} = \begin{cases} 2n-2, & \text{if } q = -1, \, n \text{ odd}, \, z^2 = -1; \text{ or } q \ne -1 \text{ and } q^n = z^2 = \pm 1, \\ 2n-1, & \text{if } q+1 \ne 0, \, z^2 \ne -1, \text{ and } q^n z^2 = 1 \text{ or } q^n z^{-2} = 1, \\ 2n, & \text{otherwise.} \end{cases}
$$

(D-E). Type $\tilde{D}_n$ $(n \ge 4)$, $\tilde{E}_n$ $(n = 6, 7, 8)$ In this part we assume that W is of type $\tilde{D}_n$ $(n \ge 4)$, $\tilde{E}_n$ $(n = 6, 7, 8)$. Then $F_1 = \Omega$. We write s_{n+1} instead of s_0.

For each $w \in c_1$, there exist unique pair $i, j \in [1, n+1]$, $\omega \in \Omega$ such that $l(s_i w) = l(w s_j) = l(w) - 1$ and $w\omega^{-1} \in W'$. We then write $w(i, j, \omega)$ instead of w. Denote $M_{n+1}(\mathbb{C}[\Omega])$ the $(n+1) \times (n+1)$ matrix ring over the group ring $\mathbb{C}[\Omega])$ and let $\pi(w(i, j, \omega)) \in M_{n+1}(\mathbb{C}[\Omega])$ be such that its (i, j)- entry is ω and other entries are 0. Then the map $w(i, j, \omega) \to \pi(w(i, j, \omega))$ defines an isomorphism of $\mathbb{C}$-algebra between $\mathbf{J}_1$ and $M_{n+1}(\mathbb{C}[\Omega])$. Since Ω is a finite abelian group, according to 5.4 we see that 12.2(a) is true.

Each $\mathbb{C}$-algebra homomorphism $f : \mathbb{C}[\Omega] \to \mathbb{C}$ gives rise to a simple representation ψ_f of $\mathbf{J}_1$:

$$\psi_f : \mathbf{J}_1 \to M_{n+1}(\mathbb{C}[\Omega]) \to M_{n+1}(C).$$

Let E_f be a simple $\mathbf{J}_1$-module providing ψ_f. As the same way in part A we know that $E_{f,q}$ has exactly one simple constituent attached to the two-sided cell c_1, denote by L_f. For different $\mathbb{C}$-algebra homomorphisms $f, f' : \mathbb{C}[\Omega]) \to \mathbb{C}$, we have $L_f \not\simeq L_{f'}$. 12.2(b) is proved.

It is easy to see that L_f is the unique simple quotient module of $E_{f,q}$. Let $v_1, v_2, ...,$ v_{n+1} be the natural base of $E_{f,q}$ and let N_f be the maximal submodule of $E_{f,q}$. then we have

Type $\tilde{E}_8$:

$$N_f = \begin{cases} < v_2 - v_3, v_3 - v_5 + v_7 - v_9 >, & \text{if } q + 1 = 0, \\ < \sum_{i=1}^{9} \alpha_i v_i >, & \text{if } (q^3 - 1)(q^5 - 1) = 0, \\ 0, & \text{otherwise,} \end{cases}$$

Where $\alpha_1 = q^{\frac{7}{2}} + q^{-\frac{7}{2}}$, $\alpha_2 = -q^2 - q^{-2} - 1$, $\alpha_3 = -q^3 - q - q^{-1} - q^{-3}$, $\alpha_i = (-1)^{\frac{10-i}{2}}(q^{\frac{10-i}{2}} + q^{\frac{10-i-1}{2}} + \cdots + q^{\frac{i+3-10}{2}} + q^{\frac{i+1-10}{2}})$ $(4 \leq i \leq 9)$. So

$$\dim L_f = \begin{cases} 7, & \text{if } q + 1 = 0, \\ 8, & \text{if } (q^3 - 1)(q^5 - 1) = 0, \\ 9, & \text{otherwise.} \end{cases}$$

Type $\tilde{E}_7$:

$$N_f = \begin{cases} < v_2 - v_5 + v_7, v_3 - v_5 + v_7 - v_8 >, & \text{if } q + 1 = 0, \\ < \sum_{i=1}^{8} \alpha_i v_i >, & \text{if } (q^3 - 1)(q^2 + 1) = 0, \\ 0, & \text{otherwise,} \end{cases}$$

Where $\alpha_1 = q^{\frac{5}{2}} + q^{-\frac{5}{2}}$, $\alpha_2 = -q - q^{-1}$, $\alpha_3 = -q^2 - 1 - q^{-2}$, $\alpha_8 = -q^3 + 1 - q^{-3}$, $\alpha_i = (-1)^{\frac{8-i}{2}}(q^{\frac{8-i}{2}} + q^{\frac{8-i-1}{2}} + \cdots + q^{\frac{i+3-8}{2}} + q^{\frac{i+1-8}{2}})$ $(4 \leq i \leq 7)$. So

$$\dim L_f = \begin{cases} 6, & \text{if } q + 1 = 0, \\ 7, & \text{if } (q^3 - 1)(q^2 + 1) = 0, \\ 8, & \text{otherwise.} \end{cases}$$

Type $\tilde{E}_6$:

$$N_f = \begin{cases} < v_1 - v_4 + v_6 + v_7 >, & \text{if } q + 1 = 0, \\ < v_1 \pm v_2 \mp v_3 - v_7, v_1 \mp v_3 \pm v_5 - v_6 >, & \text{if } (q^{\frac{1}{2}} + q^{-\frac{1}{2}}) = \pm 1, \\ < v_1 \mp 2v_2 \mp 2v_3 + 3v_4 \mp v_5 + v_6 + v_7 >, & \text{if } (q^{\frac{1}{2}} + q^{-\frac{1}{2}}) = \pm 2, \\ 0, & \text{otherwise.} \end{cases}$$

So

$$\dim L_f = \begin{cases} 5, & \text{if } (q^{\frac{1}{2}} + q^{-\frac{1}{2}}) = \pm 1, \\ 6, & \text{if } (q^{\frac{1}{2}} + q^{-\frac{1}{2}}) = 0 \text{ or } \pm 2, \\ 7, & \text{otherwise.} \end{cases}$$

Type $\tilde{D}_n$ $(n \geq 4)$:

$$N_f = \begin{cases} < v_1 - v_{n+1}, v_{n-1} - v_n, \sum_{i=1}^{\frac{n}{2}} (-1)^i v_{2i-1} >, & \text{if } q+1 = 0, \ n \text{ even,} \\ < v_1 - v_{n+1}, v_{n-1} - v_n, >, & \text{if } q+1 = 0, \ n \text{ odd,} \\ < \sum_{i=1}^{n+1} \alpha_i v_i >, & \text{if } q^{n-2} - 1 = 0, \text{ but } q+1 \neq 0, \\ 0, & \text{otherwise,} \end{cases}$$

where $\alpha_1 = \alpha_{n+1} = 1$, $\alpha_{n-1} = \alpha_n = -\frac{(-q)^{\frac{n-3}{2}} + (-q)^{\frac{3-n}{2}}}{q^{\frac{1}{2}} + q^{-\frac{1}{2}}}$, $\alpha_i = q^{\frac{i-1}{2}} + q^{\frac{1-i}{2}}$ $(2 \leq i \leq n-2)$. So

$$\dim L_f = \begin{cases} n-2, & \text{if } q+1 = 0, \ n \text{ even,} \\ n-1, & \text{if } q+1 = 0, \ n \text{ odd,} \\ n, & \text{if } q^{n-2} - 1 = 0, \text{ but } q+1 \neq 0, \\ n+1, & \text{otherwise,} \end{cases}$$

(F). Type $\tilde{F}_4$. In this part we assume that W is of type $\tilde{F}_4$. Then $F_1 = \mathbb{Z}/(2)$. We have

$c_1 = \{0, 01, 012, 0123, 01234, 01232, 012321, 0123210, 1, 10, 12, 123, 1234,$
$1232, 12321, 123210, 2, 21, 210, 23, 234, 232, 2321, 23210, 3, 32, 321, 3210, 34,$
$323, 3234, 4, 43, 432, 4321, 43210, 4323, 43234\}.$

Let $Y = \{0, 1, 2, 3, 3', 4, 4'\}$ be an F_1-set such that as F_1-sets we have $\{0\} \simeq \{1\} \simeq \{2\} \simeq F_1/F_1$ and $\{3, 3'\} \simeq \{4, 4'\} \simeq F_1$. We define a bijection

$$\pi : c_1 \overset{\sim}{\to} \text{set of irreducible } F_1\text{-v.b. on } Y \times Y \text{ (up to isomorphisms)}$$

as follows:

$0 \to 1_{(0,0)}$,	$0123210 \to \varepsilon_{(0,0)}$,	$01 \to 1_{(0,1)}$,	$012321 \to \varepsilon_{(0,1)}$,
$012 \to 1_{(0,2)}$,	$01232 \to \varepsilon_{(0,2)}$,	$0123 \to 1_{(0,3)}$,	$01234 \to 1_{(0,4)}$,
$1 \to 1_{(1,1)}$,	$12321 \to \varepsilon_{(1,1)}$,	$10 \to 1_{(1,0)}$,	$123210 \to \varepsilon_{(1,0)}$,
$12 \to 1_{(1,2)}$,	$1232 \to \varepsilon_{(1,2)}$,	$123 \to 1_{(1,3)}$,	$1234 \to 1_{(1,4)}$,
$2 \to 1_{(2,2)}$,	$232 \to \varepsilon_{(2,2)}$,	$21 \to 1_{(2,1)}$,	$2321 \to 1_{(2,1)}$,
$210 \to 1_{(2,0)}$,	$23210 \to 1_{(2,0)}$,	$23 \to 1_{(2,3)}$,	$1234 \to 1_{(2,4)}$,
$3 \to 1_{(3,3)}$,	$323 \to 1_{(3,3')}$,	$34 \to 1_{(3,4)}$,	$3234 \to 1_{(3,4')}$,
$4 \to 1_{(4,4)}$,	$43234 \to 1_{(4,4')}$,	$43 \to 1_{(4,3)}$,	$4323 \to 1_{(4,3')}$,
$32 \to 1_{(3,2)}$,	$321 \to 1_{(3,1)}$,	$3210 \to 1_{(3,0)}$,	
$432 \to 1_{(4,2)}$,	$4321 \to 1_{(4,1)}$,	$43210 \to 1_{(4,0)}$.	

Where ε is the sign representation of F_1. A direct computation shows that the bijection π satisfies (i) and (ii). 12.2(a) is proved.

Let $\alpha = \bar{1}, e = \bar{0} \in \mathbb{Z}/(2)$. By 5.4 we see that $\mathbf{J}_1$ has three simple modules (up to isomorphisms): $E_1 = E_\alpha$, $E_2 = E_e$, $E_3 = E_{e,\varepsilon}$, where ε is the sign representation of $\mathbb{Z}/(2)$. Via $\phi_{q,c_1}: \mathbf{H}_q \to \mathbf{J}_1$, they give rise to three $\mathbf{H}_q$-modules: $E_{1,q}$, $E_{2,q}$, $E_{3,q}$. Each $\mathbf{H}_q$-module $E_{p,q}$ $(p = 1, 2, 3)$ has a unique simple quotient, which is just $(\phi_q)_{*,c_1}(E_p)$. For simplicity, we denote it by L_p, then $C_i L_1 \neq 0$ and $C_i L_3 = 0$, $i = 0, 1, 2$; $C_j L_1 = 0$ and $C_j L_3 \neq 0$, $j = 3, 4$; $C_k L_2 \neq 0$, $k = 0, 1, 2, 3, 4$. Thus we have $L_1 \not\simeq L_2 \not\simeq L_3 \not\simeq L_1$. 12.2(b) is proved.

Let N_p $(p = 1, 2, 3)$ be the maximal submodule of $E_{p,q}$, then we have

$$
N_1 = \begin{cases} <v_0 - v_2>, & \text{if } q + 1 = 0, \\ <v_0 - (q^{\frac{1}{2}} + q^{-\frac{1}{2}})v_1 + v_2>, & \text{if } q^2 + 1 = 0, \\ 0, & \text{otherwise}, \end{cases}
$$

where $v_i : \{0, 1, 2\} \to \mathbb{C}$ is defined by $v_i(j) = \delta_{ij}$ $(i, j = 0, 1, 2)$. Therefore

$$
\dim L_1 = \begin{cases} 2, & \text{if } (q + 1)(q^2 + 1) = 0, \\ 3, & \text{otherwise}. \end{cases}
$$

$$
N_2 = \begin{cases} <v_0 - v_2 + v_4>, & \text{if } q + 1 = 0, \\ <\sum_0^4 \alpha_i v_i>, & \text{if } q^3 - 1 = 0, \\ 0, & \text{otherwise}, \end{cases}
$$

where $v_i : Y \to \mathbb{C}$ is defined by $v_i(j) = v_i(j') = \delta_{ij}$ $(i, j = 0, 1, 2, 3, 4)$ (we set $0 = 0', 1 = 1', 2 = 2')$, $\alpha_0 = 1$, $\alpha_1 = -q^{\frac{1}{2}} - q^{-\frac{1}{2}}$, $\alpha_2 = q + 1 + q^{-1}$, $\alpha_3 = -\frac{1}{2}(q^{\frac{3}{2}} + q^{\frac{1}{2}} + q^{-\frac{1}{2}} + q^{-\frac{3}{2}})$, $\alpha_4 = \frac{1}{2}(q + 1 + q^{-1})$. Therefore

$$
\dim L_2 = \begin{cases} 4, & \text{if } (q + 1)(q^3 - 1) = 0, \\ 5, & \text{otherwise}. \end{cases}
$$

$$
N_3 = \begin{cases} <v_3 - (q + q^{-\frac{1}{2}})v_4>, & \text{if } q^2 + q + 1 = 0, \\ 0, & \text{otherwise}, \end{cases}
$$

where $v_i : Y \to \mathbb{C}$ is defined by $v_i(j) = -v_i(j') = \delta_{ij}$ $(i = 3, 4; j = 0, 1, 2, 3, 4)$ (we set $0 = 0', 1 = 1', 2 = 2')$. Therefore

$$
\dim L_3 = \begin{cases} 1, & \text{if } q^2 + q + 1 = 0, \\ 2, & \text{otherwise}. \end{cases}
$$

Type G_2: For type $\tilde{G}_2$, Theorem 12.2 has been proved in chapter 7. We only determine here the dimensions of the simple $\mathbf{H}_q$-modules attached to the two-sided cell c_1.

By 5.4, 12.2(a) we see that $\mathbf{J}_1$ has four simple modules $E_1 = E_{\bar{e}}$, $E_2 = E_{\overline{s_0}}$, $E_3 = E_{\overline{s_0 s_1}}$, $E_4 = E_{\bar{e},\sigma}$, where $\bar{e} = \{e\}$, (e the neutral element of $\mathfrak{S}_3$), $\overline{s_0} = \{s_0, s_1, s_0 s_1 s_0\}$, $\overline{s_0 s_1} = \{s_0 s_1, s_1 s_0\}$. We have $\dim E_1 = \dim E_2 = 3$, $\dim E_3 = 2$ and $\dim E_4 = 1$. When $q + 1 \neq 0$, one verifies that $E_{i,q}$ $(1 \leq i \leq 4)$ has a unique simple quotient module L_i and $L_i = (\phi_q)_{*,c_1}(E_i)$. Moreover $L_i \not\simeq L_j$ if $i \neq j$. When $q + 1 = 0$, one can check that $E_{i,q}$ $(1 \leq i \leq 3)$ has a unique simple quotient L_i and $L_i = (\phi_q)_{*,c_1}(E_i)$. We also have $(\phi_q)_{*,c}(E_4) = 0$.

Let N_i $(i = 1, 2, 3)$ be the maximal submodule of $E_{i,q}$, then we have

$$
N_1 = \begin{cases}
< 3v_1 - v_3 >, & \text{if } q + 1 = 0, \\
< v_1 \mp 2v_2 + v_3 >, & \text{if } q = 1, \\
0, & \text{otherwise,}
\end{cases}
$$

where $v_i : Y \to \mathbb{C}$ is defined by $v_i(j) = \delta_{ij}$ $(i = 1, 2;\ 1 \leq j \leq 5)$ and $v_3(p) = 1$, $v_3(i) = 0$ $(p = 3, 4, 5;\ i = 1, 2)$. So

$$
\dim L_1 = \begin{cases}
2, & \text{if } q^2 - 1 = 0, \\
3, & \text{otherwise.}
\end{cases}
$$

$$
N_2 = \begin{cases}
< v_1 - v_3 >, & \text{if } q + 1 = 0, \\
< v_1 \mp v_2 + v_3 >, & \text{if } q^2 + 1 = 0, \\
0, & \text{otherwise,}
\end{cases}
$$

where $v_i : \{1, 2, 3\} \to \mathbb{C}$ is defined by $v_i(j) = \delta_{ij}$ $(i, j = 1, 2, 3)$. So

$$
\dim L_2 = \begin{cases}
2, & \text{if } (q^2 + 1)(q + 1) = 0, \\
3, & \text{otherwise.}
\end{cases}
$$

$$
N_3 = \begin{cases}
< v_1 - (q^{\frac{1}{2}} + q^{-\frac{1}{2}})v_2 >, & \text{if } q^2 + q + 1 = 0, \\
0, & \text{otherwise,}
\end{cases}
$$

where $v_i : \{1, 2\} \to \mathbb{C}$ is defined by $v_i(j) = \delta_{ij}$ $(i, j = 1, 2)$. So

$$
\dim L_3 = \begin{cases}
1, & \text{if } (q^2 + q + 1) = 0, \\
2, & \text{otherwise.}
\end{cases}
$$

Note that $\dim E_4 = 1$ and $E_{4,q}$ is always a simple $\mathbf{H}_q$-module (cf. 5.10).

References

[A]   D. Alvis, *The left cells of the Coxeter group of type* H_4, J. Alg. 107 (1987), 160-168.

[As]   T. Asai *et al*, *Open problems in algebraic groups*, Proc. of the Conference on "Algebraic groups and Their Representations" Held at Katata, August 29-September 3, 1983.

[BV1] D. Barbasch and D. Vogan, *Primitive ideals and orbital integrals in complex classical groups*, Math. Ann. 259 (1982), 153-199.

[BV2] D. Barbasch and D. Vogan, *Primitive ideals and orbital integrals in complex exceptional groups*, J. Alg. 80 (1983), 350-382.

[Bé1] R. Bédard, *Cells for two Coxeter groups*, Comm. Alg. 14 (1988), 1253-1286.

[Bé2] R. Bédard, *The lowest two-sided cell for an affine Weyl group*, Comm. Alg. 16 (1988), 1113-1132.

[BZ]   J. Berstein abd A.V. Zelevinsky, *Induced representations of reductive p-adic groups, I*, Ann. Sci. E.N.S. 10 (1977), 441-472.

[BB]   A. Bialynicki-Birula, *On fixed point schemes of actions of multiplicative and additive groups*, Topology 12 (1973), 99-103.

[Bo1] A. Borel, *Admissible representations of a semi-simple group over a local field with vectors fixed under an Iwahori subgroup*, Invent. Math. 35 (1976), 233-259.

[Bo2] A. Borel *et al*, *Seminar on algebraic groups and related finite groups*, LNM 131, Springer-Verlag, 1970.

[B]   N. Bourbaki, *Groups et algèbre de Lie*, Ch. 4–6, Herman, Paris, 1968.

[BM]   M. Broué and G. Malle, *Zyklotomische Heckealgebren*, Astérisque 212 (1993), 119-189.

[Br]   R.K. Brylinski, *Limits of weight spaces, Lusztig's q-analogs and fiberings of adjoint orbits*, Jour. Amer. Math. Soc. 2 (1989), 517-533.

[C1]   R.W. Carter, *Conjugacy classes in the Weyl groups*, Compositio Math. 25 (1972), 1-59.

[C2]   R.W. Carter, *Finite groups of Lie type: Conjugacy classes and complex characters*, Wiley Interscience, London, 1985.

[Ca]   P. Cartier, *Representations of p-adic groups: A survey*, Proc. Symp. Pure Math. vol. XXXIII (part I), 111-155.

[Ch]   C. Chen, *the decomposition into left cells of the affine Weyl group of type* $\tilde{D}_4$, J. Alg. 163 (1994), 692-728.

[CLP] C.De Concini, G. Lusztig, and C. Procesi, *Homology of the zero-set of a nilpotent vector field on a flag manifold*, Jour. Amer. Math. Soc. 1 (1988), 15-34.

[Cu] C.W. Curtis, *Representations of finite groups of Lie type*, Bull. Amer. Math. Soc. (N.S) 1 (1979), 721-757.

[D] J. Du, *The decomposition into cells of the affine Weyl group of type $\tilde{B}_3$*, Comm. Alg. 16 (1988), 1383-1409.

[Du] M. Duflo, *et al, Open problems in representation theory of Lie groups*, Proc. of the Conference on "Analysis on Homogeneous spaces" Held at Katata, August 25-August 30, 1986.

[F] W. Fulton, *Intersection Theory*, Springer-Verlag, 1984.

[FM] W. Fulton and R. MacPherson, *Categorical framework for the study of singular spaces*, Mem. Amer. Math. Soc. 31 (1981), no. 243.

[G1] V. Ginzburg, *Deligne-Langlands conjecture and representations of Hecke algebras*, preprint, 1985.

[G2] V. Ginzburg, *Lagrangian construction for representations of Hecke algebras*, Adv. in Math. 63 (1987), 100-112.

[G3] V. Ginzburg, *Proof of the Deligne-Langlands conjecture*, (AMS translation) Soviet Math. Dokl. Vol. 35 (1987), 304-308.

[G4] V. Ginzburg, *Geometrical aspects of representations*, Proc. Inter. Congress Math., Berkeley, 1986, vol. 1, 840-847, Amer. Math. Soc., Providence, 1987.

[GV] V. Ginzburg and E. Vasserot, *Langlands reciprocity for affine quantum groups of type A_n*, Inter. Math. Research Notices, 1993, No. 3, (in Duke Math. J., Vol. 69, No. 3 (1993)), 67-85.

[Go] M. Goresky, *Kazhdan-Lusztig polynomials for classical groups*, Northeastern University Mathematics Department [no date].

[Gu] R.K. Gupta, *Characters and the q-analog of weight multiplicity*, J. London. Math. Soc. (2) 36 (1987) 68-76.

[GU] A. Gyoja and K. Uno, *On the semisimplicity of Hecke algebras*, J. Math. Soc. Japan Vol. 41, No.1, (1989),75-79.

[HS] R. Hotta and T.A. Springer, *A specialization theorem for certain Weyl group representations and an application to the Green polynomials of Unitary group*, Invent. Math. 41 (1977), 113-127.

[Hu] J.E. Humphreys, *Reflection groups and Coxeter groups*, Cambridge studies in advanced mathematics 29, Cambridge University Press, 1990.

[I] N. Iwahori, *On the structure of a Hecke ring of a Chevalley group over a finite field*, J. Fac. Sci. Uni. Tokyo Sect. IA 10 (1964), 215-236.

[IM]   N. Iwahori and H. Matsumoto, *On some Bruhat decomposition and the struc-ture of Hecke rings of p-adic Chevalley groups*, Publ. Math. I.H.E.S. 25 (1965), 5-48.

[J1]   A. Joseph, *A characteristic variety for the primitive spectrum of a semisimple Lie algebra*, LNM 586, 102-118, Springer-Verlag.

[J2]   A. Joseph, *W-module structure in the primitive spectrum of the enveloping algebra of a semisimple algebra*, LNM 728, 116-135, Springer-Verlag.

[K]   V.G. Kac, *Infinite dimensional Lie algebras*, 2nd edition., Cambridge University Press, 1985.

[Ka1] S.I. Kato, *Irreduciblity of principal series representations for Hecke algebras of affine Type*, J. Fac. Sci. Uni. Tokyo Sect. IA 28 (1982), 929-943.

[Ka2] S.I. Kato, *Spherical functions and a q-analogue of Kostant's weight multi-plicity formula*, Invent. Math. 66 (1982), 461-468.

[Ka3] S.I. Kato, *A realization of irreducible representations of affine Weyl groups*, Indag. Math. 45 (1983), 193-201.

[KL1] K. Kazhdan and G. Lusztig, *Representations of Coxeter Groups and Hecke algebras*, Invent. Math. 53 (1979), 165–184.

[KL2] D. Kazhdan and G. Lusztig, *Schubert varieties and Poincaré duality*, in "Ge-ometry of the Laplace operator, Proc. Symp. Pure math. 34", Amer. Math. Soc., Providence, RI, 1980, pp. 185-203.

[KL3] D. Kazhdan and G. Lusztig, *Equivariant K-theory and representations of Hecke algebras, II*, Invent. Math. 80 (1985), 209–231.

[KL4] D. Kazhdan and G. Lusztig, *Proof of the Deligne-Langlands conjecture for Hecke algebras*, Invent. Math. 87 (1987), 153–215.

[Ko]   B. Kostant, *Lie algebra cohomology and the generalized Borel-Weil theorem*, Ann. of Math. 74 (1961), 329-387.

[La]   R.P. Langlands, *Problems in the theory of automorphic forms*, LNM 170, Springer-Verlag, 1970, pp. 18-86

[Lu]   D. Luna, *Slice étale*, Bull. Soc. Math. France 33 (1973), 81-105.

[L1]   G. Lusztig, *Representations of finite Chevalley groups*, C.B.M.S. Regional Conference series in Math., 39, Amer. Math. Soc., 1978.

[L2]   G. Lusztig, *Hecke algebras and Jantzen's generic decomposition patterns*, Adv. in Math. 37 (1980), 121-164.

[L3]   G. Lusztig, *On a theorem of Benson and Curtis*, J. Alg. 71 (1981), 490-498.

[L4]   G. Lusztig, *Some examples on square integrable representations of semisimple p-adic groups*, Trans of AMS, vol. 277 (1983), 623–653.

[L5] G. Lusztig, *Singularities, character formulas, and a q-analog of weight multiplicities*, Analyse et topologie sur les espeaces singuliers (II-III), Astérisque
101-102 (1983), 208-227.

[L6] G. Lusztig, *Left cells in a Weyl groups*, in "Lie Group Representation I",
LNM 1024, Springer-Verlag, Berlin, 1984, pp. 99-111.

[L7] G. Lusztig, *Characters of reductive groups over a finite field*, Ann. of Math.
Studies 107, Princeton University Press, 1984.

[L8] G. Lusztig, *The two-sided cells of the affine Weyl group of type A_n*, in "Infinite dimensional groups with applications", Springer-Verlag, New York, 1985,
pp. 275-287.

[L9] G. Lusztig, *Equivariant K-theory and representations of Hecke algebras*,
Proc. Amer. Math. Soc. 94 (1985), 337-342.

[L10] G. Lusztig, *Sur les cellules gauches des groupes de Weyl*, C.R. Acad. Paris
Sér. I Math. 302 (1986), 5-8.

[L11] G. Lusztig, *Cells in affine Weyl groups*, in "Algebraic Groups and Related
Topics", Advanced Studies in Pure Math., vol. 6, Kinokunia and North
Holland, 1985, pp. 255-287.

[L12] G. Lusztig, *Cells in affine Weyl groups II*, J. Alg. 109 (1987), 536-548.

[L13] G. Lusztig, *Cells in affine Weyl groups III*, J. Fac. Sci. Univ. Tokyo Sect.
IA Math. 34 (1987), 223-243.

[L14] G. Lusztig, *Cells in affine Weyl groups IV*, J. Fac. Sci. Univ. Tokyo Sect.
IA Math. 36 (1989) No. 2, 297-328.

[L15] G. Lusztig, *Leading coefficients of character values of Hecke algebras*, Proc.
Symp. Pure Math. vol 47 (1987), 235-262.

[L16] G. Lusztig, *Cuspidal local systems and graded Hecke algebras I*, Publ. Math.
I.H.E.S. 67 (1988), 145-202.

[L17] G. Lusztig, *Representations of affine Hecke algebras*, Astérisque 171-172
(1989), 73-84.

[L18] G. Lusztig, *Affine Hecke algebras and their graded version*, J. Amer. Math.
Soc. 2 (1989), 599-635.

[L19] G. Lusztig, *Canonical bases arising from quantized enveloping algebras*, J.
Amer. Math. Soc. 3 (1990), 447-498.

[L20] G. Lusztig, *Intersection cohomology methods in representation theory*, Proc.
Inter. Congress Math. Kyoto, 1990, vol. 1, pp. 155-174, Math. Soc. of
Japan and Springer-Verlag, 1991.

[LX] G. Lusztig and N. Xi, *Canonical left cells in affine Weyl groups*, Adv. in
Math. 72 (1988), 284-288.

[MS]   J.G.M. Mars and T.A. Springer, *Character sheaves*, Astérisque 173-174 (1989), 111-198.

[M]   H. Matsumoto, *Analyse Harmonique dans les systèmes de Tits Bornologiques de type affine*, LNM 590, Springer-Verlag, 1979.

[OV]   A.L. Onischnik and E.B. Vinberg, *Lie groups and algebraic groups*, Springer-Verlag, 1990.

[R1]   J.D. Rogawski, *On modules over the Hecke algebra of a p-adic group*, Invent. Math. 79 (1985), 443-465.

[R2]   J.D. Rogawski, *representations of GL(n) over a p-adic field with an Iwahori-fixed point*, Israel Jour. of Math. 54 (1986), 242-256.

[Sh1]   J.-Y. Shi, *Kazhdan Lusztig cells of certain affine Weyl groups*, LNM 1179, Springer-Verlag, Berlin, 1986.

[Sh2]   J.-Y. Shi, *A two-sided cell in an affine Weyl group I*, J. London Math. Soc. 36 (1987), 407-420.

[Sh3]   J.-Y. Shi, *A two-sided cell in an affine Weyl group II*, J. London Math. Soc. 37 (1988), 235-264.

[Sh4]   J.-Y. Shi, *Left cells in affine Weyl groups*, Tohoku Math. Jour. 46 (1994), 105-124.

[Sh5]   J.-Y. Shi, *Left cells in the affine Weyl groups $W_a(\tilde{D}_4)$*, to appear in Osaka J. Math. 31 (1994).

[St1]   R. Steinberg, *Regular elements of semisimple algebraic groups*, Publ. Math. I.H.E.S. 25 (1965), 49-80.

[St2]   R. Steinberg, *On a theorem of Pittie*, Topology 14 (1975), 173-177.

[T]   R. Thomason, *Equivariant algebraic versus topological K-homology Atiyah-Segal-style*, Duke Math. J. 56 (1988), 689-636.

[V1]   D. Vogan, *A generized τ-invariant for the primitive spectrum of a semisimple Lie algebra*, Math. Ann. 242 (1979), 209-224.

[V2]   D. Vogan, *Ordering in the primitive spectrum of a semisimple Lie algebra*, Math. Ann. 248 (1980), 195-203.

[X1]   N. Xi, *Induced cells*, Proc. of Amer. Math. Soc. 108 (1990), 25-29.

[X2]   N. Xi, *The based ring of the lowest two-sided cell of an affine Weyl group*, J. Alg. 134 (1990), 356-368.

[X3]   N. Xi, *The based ring of the lowest two-sided cell of an affine Weyl group, II*, Ann. Sci. E.N.S. 27 (1994), 47-61.

[X4] N. Xi, *The based ring of cells in affine Weyl groups of type* $\tilde{G}_2$, $\tilde{B}_2$, preprint, 1992.

[X5] N. Xi, *Some simple modules of affine Hecke algebras*, preprint, 1992.

[Z]   A.V. Zelevinsky, *Induced representations of reductive p-adic groups, II. On irreducible representations of* GL_n, Ann. Sci. E.N.S. 13 (1980), 165-210.

Notation

Index

Vol. 1547: P. Harmand, D. Werner, W. Werner, M-ideals in Banach Spaces and Banach Algebras. VIII, 387 pages. 1993.

Vol. 1548: T. Urabe, Dynkin Graphs and Quadrilateral Singularities. VI, 233 pages. 1993.

Vol. 1549: G. Vainikko, Multidimensional Weakly Singular Integral Equations. XI, 159 pages. 1993.

Vol. 1550: A. A. Gonchar, E. B. Saff (Eds.), Methods of Approximation Theory in Complex Analysis and Mathematical Physics IV, 222 pages, 1993.

Vol. 1551: L. Arkeryd, P. L. Lions, P.A. Markowich, S.R. S. Varadhan. Nonequilibrium Problems in Many-Particle Systems. Montecatini, 1992. Editors: C. Cercignani, M. Pulvirenti. VII, 158 pages 1993.

Vol. 1552: J. Hilgert, K.-H. Neeb, Lie Semigroups and their Applications. XII, 315 pages. 1993.

Vol. 1553: J.-L- Colliot-Thélène, J. Kato, P. Vojta. Arithmetic Algebraic Geometry. Trento, 1991. Editor: E. Ballico. VII, 223 pages. 1993.

Vol. 1554: A. K. Lenstra, H. W. Lenstra, Jr. (Eds.), The Development of the Number Field Sieve. VIII, 131 pages. 1993.

Vol. 1555: O. Liess, Conical Refraction and Higher Microlocalization. X, 389 pages. 1993.

Vol. 1556: S. B. Kuksin, Nearly Integrable Infinite-Dimensional Hamiltonian Systems. XXVII, 101 pages. 1993.

Vol. 1557: J. Azéma, P. A. Meyer, M. Yor (Eds.), Séminaire de Probabilités XXVII. VI, 327 pages. 1993.

Vol. 1558: T. J. Bridges, J. E. Furter, Singularity Theory and Equivariant Symplectic Maps. VI, 226 pages. 1993.

Vol. 1559: V. G. Sprindžuk, Classical Diophantine Equations. XII, 228 pages. 1993.

Vol. 1560: T. Bartsch, Topological Methods for Variational Problems with Symmetries. X, 152 pages. 1993.

Vol. 1561: I. S. Molchanov, Limit Theorems for Unions of Random Closed Sets. X, 157 pages. 1993.

Vol. 1562: G. Harder, Eisensteinkohomologie und die Konstruktion gemischter Motive. XX, 184 pages. 1993.

Vol. 1563: E. Fabes, M. Fukushima, L. Gross, C. Kenig, M. Röckner, D. W. Stroock, Dirichlet Forms. Varenna, 1992. Editors: G. Dell'Antonio, U. Mosco. VII, 245 pages. 1993.

Vol. 1564: J. Jorgenson, S. Lang, Basic Analysis of Regularized Series and Products. IX, 122 pages. 1993.

Vol. 1565: L. Boutet de Monvel, C. De Concini, C. Procesi, P. Schapira, M. Vergne. D-modules, Representation Theory, and Quantum Groups. Venezia, 1992. Editors: G. Zampieri, A. D'Agnolo. VII, 217 pages. 1993.

Vol. 1566: B. Edixhoven, J.-H. Evertse (Eds.), Diophantine Approximation and Abelian Varieties. XIII, 127 pages. 1993.

Vol. 1567: R. L. Dobrushin, S. Kusuoka, Statistical Mechanics and Fractals. VII, 98 pages. 1993.

Vol. 1568: F. Weisz, Martingale Hardy Spaces and their Application in Fourier Analysis. VIII, 217 pages. 1994.

Vol. 1569: V. Totik, Weighted Approximation with Varying Weight. VI, 117 pages. 1994.

Vol. 1570: R. deLaubenfels, Existence Families, Functional Calculi and Evolution Equations. XV, 234 pages. 1994.

Vol. 1571: S. Yu. Pilyugin, The Space of Dynamical Systems with the C^0-Topology. X, 188 pages. 1994.

Vol. 1572: L. Göttsche, Hilbert Schemes of Zero-Dimensional Subschemes of Smooth Varieties. IX, 196 pages. 1994.

Vol. 1573: V. P. Havin, N. K. Nikolski (Eds.), Linear and Complex Analysis – Problem Book 3 – Part I. XXII, 489 pages. 1994.

Vol. 1574: V. P. Havin, N. K. Nikolski (Eds.), Linear and Complex Analysis – Problem Book 3 – Part II. XXII, 507 pages. 1994.

Vol. 1575: M. Mitrea, Clifford Wavelets, Singular Integrals, and Hardy Spaces. XI, 116 pages. 1994.

Vol. 1576: K. Kitahara, Spaces of Approximating Functions with Haar-Like Conditions. X, 110 pages. 1994.

Vol. 1577: N. Obata, White Noise Calculus and Fock Space. X, 183 pages. 1994.

Vol. 1578: J. Bernstein, V. Lunts, Equivariant Sheaves and Functors. V, 139 pages. 1994.

Vol. 1579: N. Kazamaki, Continuous Exponential Martingales and BMO. VII, 91 pages. 1994.

Vol. 1580: M. Milman, Extrapolation and Optimal Decompositions with Applications to Analysis. XI, 161 pages. 1994.

Vol. 1581: D. Bakry, R. D. Gill, S. A. Molchanov, Lectures on Probability Theory. Editor: P. Bernard. VIII, 420 pages. 1994.

Vol. 1582: W. Balser, From Divergent Power Series to Analytic Functions. X, 108 pages. 1994.

Vol. 1583: J. Azéma, P. A. Meyer, M. Yor (Eds.), Séminaire de Probabilités XXVIII. VI, 334 pages. 1994.

Vol. 1584: M. Brokate, N. Kenmochi, I. Müller, J. F. Rodriguez, C. Verdi. Editor: A. Visintin. Phase Transitions and Hysteresis. Montecatini Terme, 1993. VII. 291 pages. 1994.

Vol. 1585: G. Frey, On Artin's Conjecture for Odd 2-dimensional Representations. VII, 148 pages. 1994.

Vol. 1586: R. Nillsen, Difference Spaces and Invariant Linear Forms. ••, ••• pages. 1994.

Vol. 1587: N. Xi, Representations of Affine Hecke Algebras. VIII, 137 pages. 1994.

Vol. 1589: J. Bellissard, M. Degli Esposti, G. Forni, S. Graffi, S. Isola, J. N. Mather, Transition to Chaos in Classical and Quantum Mechanics. Montecatini, 1991. Editor: S. Graffi. VII, 192 pages. 1994.